The Sun-Climate Connection Over the Last Millennium
Facts and Questions

Authored By

Maxim Ogurtsov

*Ioffe PhTI, Laboratory of Cosmic Ray,
St. Petersburg, 194 021, Polytechnicheskaya 26; GAO
RAN, Pulkovo, Russia*

Risto Jalkanen, Markus Lindholm

*Metla, Rovaniemi Research Unit, P.O. Box 16, FI
96301 Rovaniemi, Finland*

Svetlana Veretenenko

*Ioffe PhTI and St.Petersburg State University,
St. Petersburg, Russia*

CONTENTS

contd…..

contd.....

contd.....

Foreword

Climate change, variations of solar irradiance and physical state of the circumterrestrial space have significant impact on many kinds of human activity and the natural variability. The history of mankind knows severe examples of the vulnerability of human society of these dependences. The presence of a link between the Sun's activity, galactic cosmic ray (GCR) flux, effectively modulated by Sun, and the Earth's climate has been claimed in many works. Consequently, the possible solar contribution to the global warming of the last century is at present actively debated. However, our modern knowledge about the past of the phenomena is quite insufficient since direct instrumental records are rather short. Analysis of high-resolution solar and climatic long-term proxies makes it possible to fill the gaps in our knowledge.

The eBook is written in frame of a multi-disciplinary approach and collects efforts of specialists in different scientific fields, including solar-terrestrial relations, paleoastrophysics, paleoclimatology, dendroclimatology. It describes well the most up-to-date methods of solar paleoastrophysics and paleoclimatology and accumulates a lot of information about solar and climatic variability over time scales from decades to millennia. It contains both detailed survey of experimental results, testifying reality of solar-climatic relationship, and discussion of possible physical mechanisms providing a link between the Sun's activity and the lower atmosphere. The eBook includes some interesting results not well known outside Russia *e.g.* published only in Russian or obtained by scientists who work out of the Russian (Soviet) Academy system.

The eBook emphasizes necessity of the complex approach to the problems of solar-terrestrial and solar-climatic relationship. It will be useful for researchers which work in solar-terrestrial physics, climatology, geophysics, astronomy, for post-graduate students and students of older years.

V.N. Obridko
Institute of Terrestrial Magnestism
Russia

Preface

Solar activity is a phenomenon very important for the mankind existence. Presence of link between Sun's activity, galactic cosmic ray (GCR), effectively modulated by Sun, and terrestrial climate has been claimed by many authors and possible solar contribution to the global warming of the last century is actively debated. Clarification of the nature and mechanism of global warming will supply us information about the climate behavior during at least few future decades and, hence, is of a great importance for all the humankind.

However, many problems still have to be solved. If solar and GCR contribution to global warming really exists, what is its extent? What are the physical mechanisms, providing relationship between space weather, solar activity, and climate? What is greenhouse contribution in global warming? What is the contribution of activity of volcanoes in long-term temperature variation? Does geomagnetic dipole field affect the Earth's climate? The answer to these questions needs information about the past of space weather, solar activity, solar-terrestrial interactions, climate and geomagnetic dipole field over long time scale. However, the currently available information about past of these phenomena is not precise enough. Really the instrumental temperature time series usually cover no more than last 100-150 years. Different records of solar activity also are short. The longest of them – group sunspot numbers – started since A.D. 1610. Accurate observations of many space weather parameters cover less than 55 last years. That is why our knowledge about long-term (decadal, centennial and millennial) variability of the solar and terrestrial phenomena is very limited and many features of mechanism, linking space weather, the Sun's activity, volcanic aerosol loading and state of the terrestrial atmosphere, are still need to be clarified. This information, in turn, can be obtained only with using proxies, such as: (a) historical data (catalogues of ancient auroral and sunspot observations)? (b) data on abundance of cosmogenic isotopes and ions (^{14}C, ^{10}Be, NO_3^-, SO_4^-) in natural archives. These proxies are yielded by paleoastrophysics and can span more than 10 000 past years. Modern paleoclimatology provides irreplaceable tools in constructing increasingly confident histories or chronologies of past climatic changes using natural archives. High- resolution paleodata are particularly perspective. Dendroclimatological reconstructions, which use tree-rings as climate proxies, have the advantage of annual resolution. Stable isotope data (^{18}O, D) make it possible to investigate climate variability over much longer time scales often with annual resolution too.

Geothermal temperature reconstructions, historical documentary records, paleobotanic data, and melt layer thickness are other sources of information used by paleoclimatology.

Methods of paleoastrophysics and paleoclimatology have been intensively developed during the last 20 years. Therefore, a lot of new information has been accumulated, including many reconstructions of sunspot numbers, solar flares, global and regional temperature covering time intervals from present to 10,000 years backwards along the timeline. Recent advances in paleoclimatology and solar paleoastrophysics have made it possible to study the evolution of climate and different manifestations of the Sun's activity over long (from multi-decadal to multi-centennial) time scales, which have previously been inaccessible. We describe a huge massive of the new results, obtained in these fields by many workers, including the authors of the eBook and provide a systematization of these data, some of which are rather controversial. Modern progress in paleosciences makes it also possible to test possible solar-climate connection over the long time scale that is of a great importance. Actually, all the physical mechanisms of a link between solar activity and lower atmosphere were carried out using rather short (less than 10-20 years) experimental data sets. Examination of suitability of the known solar-climate mechanisms for explanation of the much longer phenomena, performed in the eBook, brings new evidences of the reality of solar-climate link. We provide also a summary of present understanding in solar activity evolution, solar-terrestrial relationship and the possible mechanisms determining the reaction of the terrestrial climate system to the Sun-cosmic factors. Possible scenarios of the future climate evolution are considered. A discussion of the accuracy and reliability of the available solar and climate proxies is included. We believe that the eBook will not only help the readers in developing a better understanding of the solar and climatic variability and solar-terrestrial interactions but will emphasize the necessity of further systematic and intensive study in all these areas.

ACKNOWLEDGEMENTS

M.G. Ogurtsov expresses his thanks to the exchange program between the Russian and Finnish Academies (project No. 16), to the program of the Presidium of RAS No 22, and to RFBR grants 13-02-00277, 13-02-00783 for financial support. The authors are also thankful to the anonymous referees whose recommendation allowed improve the eBook substantially.

CONFLICT OF INTEREST

The author confirms that this eBook content has no conflict of interest.

Maxim Ogurtsov
Ioffe PhTI, Laboratory of Cosmic Ray
St. Petersburg, 194 021
Polytechnicheskaya 26
and
GAO RAN, Pulkovo
Russia
E-mail: maxim.ogurtsov@mail.ioffe.ru

Risto Jalkanen, Markus Lindholm
Metla, Rovaniemi Research Unit
P.O. Box 16, FI 96301 Rovaniemi
Finland

and

Svetlana Veretenenko
Ioffe PhTI and St. Petersburg State University
St. Petersburg
Russia

CHAPTER 1

Introduction

Abstract: Some general qualities of the Sun are described. Definitions of solar and geomagnetic activity, space weather as well as climate are introduced. Highlights of the history of solar-climate research since the beginning of the 17th century are shortly discussed. In addition, the main problems and shortcomings of modern helioclimatology are presented.

Keywords: Climate, helioclimatology, solar activity, space weather.

1.1. The Sun and Heliosphere

The Sun - a giant plasma sphere - is our closest star and the principal source of energy for the Earth Table **1.1**. The energy radiating from the Sun affects significantly almost all terrestrial processes (both physical and biological).

Table 1.1. General features of the Sun.

Characteristic	Value
Equatorial radius	6.96×10^{10} cm
Mass	1.99×10^{33} g
Angular diameter	32′30″(1919 arc seconds)
Luminosity	3.84×10^{26} W
Mean density	1.41 g×cm^{-3}
Solar irradiance at the Earth's orbit	1367 W×m^{-2}
Effective surface temperature	5780°K

The solar atmosphere consists of the photosphere and the chromosphere. Visible sunlight is emitted from the ***photosphere,*** which is a few hundreds of kilometres thick and has a temperature of 5800-6300°K (see also Fig. **1.1**). The ***chromosphere*** is an irregular layer above the photosphere where temperature increases from ca 6300°K to about 20,000°K. The chromosphere is 1800-3000 km thick. Its visible image can be observed during solar eclipses (see also Fig. **1.1**).

Solar activity is a number of non-stationary physical processes and phenomena in the solar atmosphere related to the changes in the Sun's magnetic fields. Solar

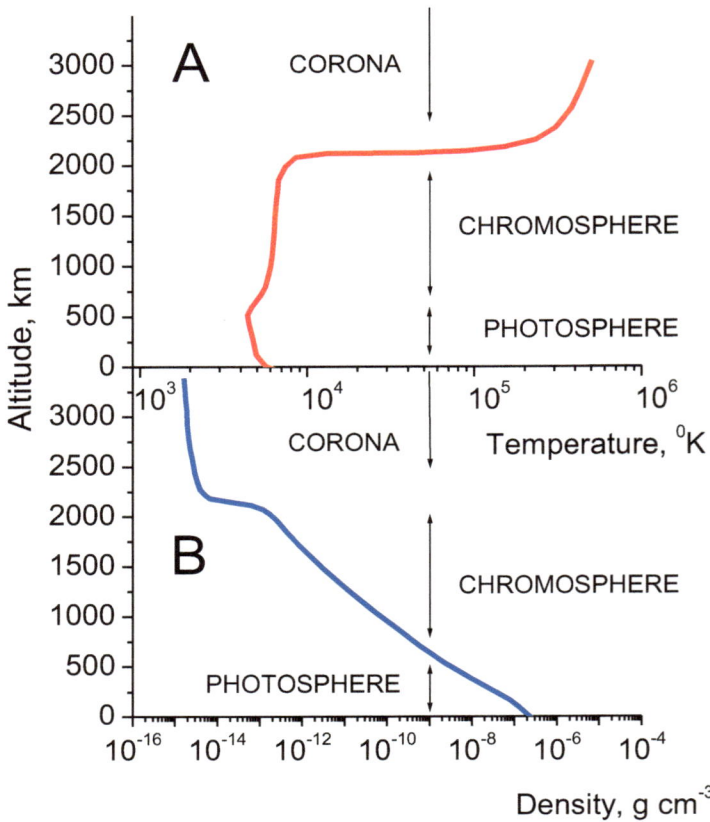

Fig. (1.1). Temperature and density distribution in the Sun's atmosphere.

activity (SA) manifests itself of by appearance and further time evolution of sunspots, faculae, flocculae, protuberances, solar flares, fluxes of solar corpuscular as well as electromagnetic radiation.

Sunspots are the most evident signs of solar activity. They are occasional features of the photosphere that emerge as visible dark dots contrasting with the surrounding bright area. They are caused by strong magnetic fields, which reduce convection, forming areas of lower surface temperature.

Heliosphere is a region allocated in interstellar space around the Sun where solar wind and solar magnetic fields dominate. Measurements by the equipment of the spacecraft *Voyager* showed that *heliopause* - the border separating solar wind from the plasma of the interstellar environment - is situated at a distance of

100-150 astronomic units (1 astronomic unit is 149.61×10^6 km - an average distance between the Sun and the Earth) from the Sun.

1.2. The Earth and Circumterrestrial Space

The Earth has a mean radius of 6.37×10^8 cm and a surface area of 5.1×10^{18} cm^2. Oceans and seas cover 71% of the Earth's surface, *i.e.* 3.62×10^{18} cm^2. Terrestrial atmosphere consists of the troposphere (up to 10-17 km), the stratosphere (10-17 to 50 km), the mesosphere (50-80 km) and the thermosphere (above 80 km) [1]. Its chemical composition is 78% of N_2, 21% of O_2 and a number of trace gases (CO_2, noble gases). Below the height of 80 km the atmosphere is neutral but above this level the relative number of atmospheric ions increases, forming the ionosphere (80-1000 km). The ionosphere usually includes layers D (60-90 km), E (90-120 km) and F layer (200-500 km). The Earth has an internal dipole magnetic moment of 8×10^{22} A m^2 which produces a magnetic field, the strength of which is about 3.1×10^{-5} T (0.3 Gauss) at the equator on the Earth's surface. This field decreases with radial distance as r^{-3}.

Magnetosphere is a circumterraneous region where the motion of charged particles is driven by the Earth's magnetic field. It has a complex structure determined by the interaction between the magnetic field of the Earth, the solar wind, and the upper atmosphere plasma. The lower part of magnetosphere lies at an altitude of a few hundreds of kilometres while its tail has a size up to 40 Earth's radii [2].

Geomagnetic activity is the disturbances of the geomagnetic field caused by variations in the magnetosphere-ionosphere current system. It is closely connected to variations in solar particle flux. The main manifestations of geomagnetic activity are geomagnetic storms, substorms and auroral phenomena. A strong magnetic storm can change the global field by 5×10^{-7} T or more [3].

Solar-terrestrial physics or *heliogeophysics* is a number of scientific disciplines which study the phenomena and processes at the Sun and their manifestations in circumterraneous space and the Earth's atmosphere and magnetosphere. It is evidently a complex and extensive field and thus incorporates solar physics, astrophysics, cosmic physics, geomagnetism, physics of the atmosphere, meteorology, climatology and possibly some other scientific branches of science [3]. Potential effect of solar activity on the Earth's climate is the key interest in *helioclimatology*. As to the definition of 'climate', it is convenient to define it using the concept of weather.

1.3. Weather and Climate

Atmospheric weather is the physical state of the atmosphere in a given point of the Earth at an appointed time or time interval, which is described by a number of meteorological parameters. Meteorological parameters are the general features of the atmosphere's physical condition including temperature, pressure, quality and concentration of water droplets, vapour, aerosols, speed and direction of wind. The changes of weather are fluctuations of meteorological parameters connected with pure atmospheric processes. Thus it is reasonable to determine the length of the timescale of weather phenomena as a lifetime of atmospheric instabilities. Baroclinic instability - the most general atmospheric instability connected with cyclones and anticyclones - has a lifetime of close to one week. Internal atmospheric oscillations over longer timescales - the oscillation of Madden-Julian - has been revealed recently. This oscillation is a perturbation of the pressure fields and precipitation over vast areas of the Indian and Pacific Oceans, which continues for 20-80 days and is repeated a few times per year. That is why it is reasonable to consider a time interval from one day to a few months as a characteristic scale of weather changes [4].

Climate is a statistical weather regime, which is determined by meteorological parameters averaged over some time interval. Classical climatology regards 30 years as the reference period for computing climate. In the present work we consider meteorological processes with timescales of longer than one year as climatic. This definition does not agree with the classification adopted in classical climatology but has a clear physical basis. Actually, besides internal oscillations the atmosphere could have forced oscillations, linked to the heat exchange with the main heat reservoirs (ocean, surface with biota, cryosphere) and as a result be dependent on the processes in these reservoirs. Typical timescales of these processes are much longer, *e.g.* the lifetime of temperature anomalies of the ocean surface (upper 75-200 m) reaches 5 months [4]. Characteristic times of the fluctuations of the thermal regime of surface, cryosphere and deep ocean ranges vary from a few years up to many centuries. Taking these considerations into account the timescale of one year was selected here as a border separating weather from climate. Thus all variations of meteorological parameters with timescales shorter than one year are considered as weather fluctuations, while variability with periods longer than one year (*e.g.* decadal and bidecadal) are considered as climatic.

Space weather is a physical state of the near-Earth space (upper atmosphere, magnetosphere) at a given moment or time interval, which is described by a number of heliogeophysical parameters. The term 'space weather' likely appeared first in the work of Gold [5]. The intensity of solar electromagnetic and corpuscular radiation, the fluxes of solar cosmic rays (SCR), the velocity and density of solar wind particles, the rate of ion generation in the atmosphere and the intensity of geomagnetic fluctuations are the main heliogeophysical parameters which form space weather. Space weather parameters averaged over the time intervals longer than one year determine *space climate*. Space climate as well as space weather both closely linked to solar activity.

B. Balliani probably was the first who assumed a connection between solar activity and terrestrial climate. In his letter to Galileo Balliani noted that sunspots can be considered as coolers of the Sun and hence of the Earth [3]. Since then the search for a possible link between solar activity and climate has attracted many scientists. Over the recent few decades helioclimatological investigations have become an important part of solar-terrestrial physics. These years have been a period of intensive theoretical and experimental investigation of a whole range of solar, solar-terrestrial and solar-heliosphere processes and phenomena, both theoretical and experimental, performed using both groundbased, baloon and space-borne experiments. This attention to climate, solar activity and space weather is not surprising. Indeed, the impact of changes in the Earth's climate (both global and regional) on many aspects of the social and economic activity of civilization is undoubtedly great - examples range from the consequences of the famine in Sahel (south Sahara) in 1968-1972 to the downfall of South American economies due to El Niño in 1998. Adverse conditions in the near-Earth space can, for example substantially deteriorate radio communications, disturb radio navigation and damage satellite equipment [6-9]. The most severe space weather events can damage electronic power grids; the extreme flare in August 1972 probably would have killed an astronaut in space [10].

1.4. Problems and Guidelines for Further Research

The projected problems related to global warming have already become an important issue of global politics and societal planning. Increasing our knowledge about the nature and mechanism of global warming has been essential in understanding and predicting climate behaviour in a few future decades at least, and hence is of great importance for all humanity. However, a lot of questions still remain and many problems have yet to be solved, *e.g.* Is global warming a result of the greenhouse effect entirely or do other factors, such as the increase of solar irradiance and the decrease of galactic cosmic rays (GCR) intensity, also contribute? If solar and GCR contribution to global warming actually exists, what is its extent? What is the physical mechanism providing a connection between solar activity, space weather and climate? Does the change in the geomagnetic dipole field affect the Earth's climate? What is the role of natural climatic variability in the long-term climatic change? What is the role of regional anthropogenic factors like urbanization and land-use changes? The answers to these and many other questions need detailed information about the past of climate, volcanic activity, solar activity, space weather and geomagnetic dipole field over time scales from centuries to millennia.

Unfortunately, current knowledge about the past of these phenomena is rather poor and has a lot of gaps. Indeed, systematic meteorological observation started around 1850 and thus instrumental temperature records in general cover only the last 100-150 years. More or less precise data on volcanic activity cover the same time interval. In addition, instrumental series of different parameters of the Sun's activity are rather short. The longest of the available records - group sunspot numbers - started in AD 1610. The longest of the geomagnetic time series - the C9 index of the Pavlovsk-Vojejkovo geomagnetic observatory - started in AD 1841. Moreover, accurate and thorough observations of solar flares, active study of GCR using neutron monitors and systematic satellite surveillance of various space weather parameters cover less than 55 years. This is why our knowledge of the long-term (decadal, centennial and millennial) variability of these solar and terrestrial phenomena is very limited and many key features, as well as details of the mechanisms, linking space weather, solar activity, volcanic aerosol loading and the state of the Earth's atmosphere still need to be clarified. Despite a long period of research we still do not have unambiguous and definitely established proof of actual influence of the Sun's activity on terrestrial climate. Thus climatologists often neglect the possible effect of solar activity on climate. For example, in the report of the IPCC [11] the possible solar contribution to global warming was assessed as about 8% of the anthropogenic one. The main claims of climatologists regarding helioclimatology can be summarized to the following (see [12-15]):

1)   Solar-climatic links reported in many works are temporally and spatially unstable. They appear and disappear, signs of correlations change *etc*. Furthermore, possible connections between solar activity and weather and climatic phenomena are sometimes studied in a careless manner. Thus it is claimed that there are no reasons to talk about a causal link between the Sun's activity and climate at all.

2)   The mechanism of the possible influence of solar activity on the troposphere is unknown.

3)   Forecasts of solar activity for time intervals of interest to climatology (decades and longer) are practically unreliable or nearly impossible.

These critical points are certainly based on some reasonable grounds. However, it is worth noting that our knowledge of many important heliogeophysical processes (the fluxes of energetic cosmic particles, the strength of interplanetary magnetic fields, solar wind velocity, many geomagnetic indexes) covers no more than 3-5 solar cycles. Thus, the gap in our current understanding of the long-term evolution of solar-terrestrial phenomena could be the cause for the claimed shortcomings of helioclimatology. The necessary information can be gained by means of proxies provided by solar palaeoastrophysics and palaeoclimatology. The last three decades has seen huge development of solar palaeoastrophysics - the science, which methods enable to reconstruct different parameters of solar activity in the pre-instrumental period. Palaeoclimatology - the scientific discipline aimed at the research of the climate of the past - has also been actively developed during the same period. Of course, analysing paleodata is not an easy task. They are more uncertain than experimental data and the uncertainty increases with time. A serious problem is connected with the fact that the majority of paleoproxies are non-stationary and many of them contain strong noise components. Moreover, noises that contaminate climatic signal are often "colored" and non-additive. Hence, an analysis of the time structure of available datasets to reveal possible interrelation between them is not an easy and standard problem, but require novel statistical approaches. Combined research efforts of different kinds of proxies is necessary for the extraction of reliable information about the past of solar activity and climate in order to better understand their long-term evolution. This calls for joint efforts of specialists in different scientific fields. The increasing importance of knowledge about our environmental history for many aspects of the wellbeing of mankind makes this work valuable. Despite some difficulties, new achievements of palaeoastrophysics and palaeoclimatology allow us to study the evolution of solar activity and climate and examine their possible interrelations over the long timescales that were inaccessible earlier. This is the main task of this monograph. Systematization of a large volume of information obtained over the recent decades is our second aim. The monograph includes a survey of results of Soviet-Russian researchers not often known for readers outside Russia.

CHAPTER 2

Solar Activity and the Solar-Terrestrial Connection

Abstract: The main manifestations of solar activity, such as sunspots, floccules and faculae are described. A short history of research on the Sun's activity from Galileo and Scheiner to modern times is presented. The main instrumental solar indexes - Wolf number, group sunspot number, sunspot area - and their general statistical laws (Gnevyshev-Ohl rule, Waldmeier rule *etc.*) are described. Up-to-date modern knowledge of major solar periodicities (cycles of Schwabe, Hale, Gleissberg *etc.*) and global extremes of solar activity (Maunder and Spörer minima, Medieval maximum *etc.*) are discussed. Moreover, some major phenomena related to space weather (solar wind, aurora borealis *etc.*) and other terrestrial manifestations of solar activity are considered. Modern methods of statistical analysis of non-stationary time series, including Fourier, wavelet, and singular spectrum analyses, long-range correlation analysis and multi-fractal approach are introduced. A method used for the evaluation of significance in details of wavelet spectra is presented. Furthermore, some enduring problems are shortly described.

Keywords: Solar activity, solar-terrestrial relationship, space weather.

2.1. General Manifestations of Solar Activity

Sunspots are magnetic structures that emerge dark on the solar disc. A temporary concentration in the magnetic field inhibits convection of hot matter from the Sun's interior, resulting in a cooler, darker area on the photosphere. Each sunspot is consists of a dark central core, the umbra, and a more bright halo, the penumbra. The existence of a penumbra differentiates sunspots from the usually smaller *pores*. The area of an umbra is 0.17-0.25 of the full sunspot area [16].

The diameter of a usual sunspot is 22,000-29,000 km (30-40″). The corresponding area is 120-160 $\times 10^{-6}$ of the visible hemisphere or Micro Solar Hemisphere (MSH = 3.32×10^{16} cm^2) [17]. The largest sunspots can reach size of 70,000 km or more whereas the smallest sunspots are about 3,500 km in diameter [18]. The diameter of the largest pores can reach 7000 km and thus exceed the diameter of the smallest sunspots. The complex of a number of sunspots forms a *sunspot group*. The magnetic field averaged over a sunspot is approximately 1000-1500 G. The maximum of a sunspot magnetic field in G (1 Gauss = 10^{-4} Tesla) could be evaluated using formula [19]:

Maxim Ogurtsov, Risto Jalkanen, Markus Lindholm and Svetlana Veretenenko

$$B_m = \frac{3700 \ A}{A + 66},$$ (2.1)

where A is the area of the spot in MSH.

The lifetime of a sunspot tends to increases linearly with its area, following the so-called Gnevyshev-Waldmeier rule:

$$T = \frac{A}{W},$$ (2.2)

where T is lifetime in days and $W \cong 10$ MSH day^{-1}. Thus the mean lifetime of a typical sunspot is of a few weeks while some of them can exist for several months. The total solar contrast is defined as:

$$\alpha = 1 - \frac{I_s}{I_{bgr}},$$ (2.3)

where I_s is the wavelength-integrated sunspot brightness and I_{bgr} is the same for the surrounding background photosphere [18, 19]. Temperature, magnetic field and contrast α are different in umbra and penumbra (Table **2.1**). The magnetic tubes providing sunspots can penetrate deep in the solar interior up to 2000-4000 km [20]. The clearly manifested increase and decrease of sunspot numbers over an approximate 11-year period is the key feature of sunspot activity. Sunspots appear inside the bands reaching up to 30° on each side of the solar equator and usually form groups. The latitudes of the sunspots vary in the course of a solar cycle. The first sunspots emerge close to latitudes 30°, while the last sunspots of a cycle appear close to the solar equator.

Faculae are long-living (typical lifetime is 3 times longer than that of sunspots) bright areas, usually situated near sunspots. Facular fields are composed of magnetic elements, small (<300 km diameter) flux tubes. The size of an individual facula can reach 10,000 km. Faculae usually occur on the photosphere, but sometimes can extend upwards into the chromosphere. Their temperature is higher than that of the surrounding photosphere and thus they are usually brighter (Table **2.1**). The area on the Sun covered by faculae follows the 11-year cycle. Typically it is 15-20 times larger than the sunspot area. Therefore the luminosity of the Sun correlates positively with the number of sunspots.

Table 2.1. Physical features of sunspots and faculae.

		Effective Temperature, °K	Magnetic Field, G	Total Solar Contrast
Sunspot	Umbra	3800-4800	1800-3700	0.06-0.25
	Penumbra	5400-5800	700-2000	0.50-0.80
Facula		6100	200-400	1.10-1.45

A sunspot group together with the surrounding faculae forms an active complex. The magnetic field outside such an active complex in general is less than 10 G (10^{-3} T) [16].

A *solar flare* is a powerful explosive event in the Sun's chromosphere. The power of a typical flare is typically in the order of 10^{27} erg s^{-1} and the total energy released during a flare is 10^{29}-3×10^{32} erg (*e.g.*, [21]). The energy of a solar flare can be estimated by means of the formula of Shibata *et al.* [22]:

$$E_{flare} = 7\times10^{32}\left(\frac{f}{0.1}\right)\left(\frac{B}{10^3 G}\right)^2\left(\frac{A\ /3.04\times10^{22}\,cm^2}{0.001}\right)^{3/2}, \qquad (2.4)$$

where E_{flare} in erg, f ($\delta 1.0$) is a fraction of the magnetic energy which can be released as flare energy, B is the strength magnetic field in Gauss, A is the area of the sunspot in cm^2, and 3.04×10^{22} cm^2 is a half of the solar surface.

Solar flares can emit energy in a variety of forms, including both accelerated particles and electromagnetic radiation released in spectral ranges from gamma rays and X-rays ($\lambda = 2\times10^{-11}$ cm) to radio waves ($\lambda = 10^6$ cm). Solar flares may have three stages: (a) precursor phase, (b) impulsive phase and (c) decay phase. The precursor or pre-flare stage can be detected by soft X-ray and H$_\alpha$ emission. This phase is observed in very large flares. In the second - impulsive or flash stage - most of the flare energy is released and particles are accelerated to high - 1 MeV or more - energies (1 ev = 1.6×10^{-12} erg). During an impulsive phase the harder part of the electromagnetic spectrum (gamma rays and X-rays) is emitted. In the third phase, a steady stage of decay manifests itself with soft X-ray, H$_\alpha$ and radio wave emission. The duration of these phases can range from several seconds to several hours. The majority of solar flares occur in regions of enhanced and complex magnetic fields (active regions) [23]. The phenomenon of magnetic reconnection plays a key role in the energy release produced by a solar flare. The mean area of a solar flare is about 160 MSH and its lifespan (duration) lays in the range from minutes to hours. Solar flares are classified in relation to their X-ray flux in the wavelength range from 1 to 8 Ångstroms near Earth. They may belong to the following categories: X-class flares having a maximum flux of order 10^{-4} W m^{-2}, M-class flares (a peak flux of order 10^{-5}-10^{-4} W m^{-2}), C-class flares (a maximum flux of order 10^{-6}-10^{-5} W m^{-2}) and B-class flares (a maximum flux less than 10^{-6} W m^{-2}). Large flares which produce fluxes of relativistic electrons and high-energy protons are called *proton flares*. The energy of protons in these exceptional flares can reach 10 GeV or more. *White-light flares* are are prominent events which are visible in white light. Such flares are usually strong emitters of X-rays, radio waves and accelerated particles.

2.2. A Brief History of Solar Activity Research

Naked-eye observations of sunspots have a long history. In particular, ancient Oriental (Chinese and Korean) observers obtained detailed, although scarce and deficient, data sets going back over the two millennia [24, 25].

Kepler (1571-1630) probably made the first instrumental sunspot observation in 1607 using a camera-obscura [26]. However, Kepler attributed the observed phenomenon to a transit of Mercury. Sunspots were discovered in 1610-1611 by Galileo (1564-1642), Scheiner (1575-1650), Harriot (1560-1621) and Fabricius (1587-1615) who observed the Sun by means of a telescope. Although telescopic observations of sunspots started relatively early, their quality throughout the 17th century was rather poor. First telescopes had magnification of no more than 40x. In addition, they also suffered from poor quality glass and chromatic aberration [27]. Towards the mid century the astronomical optics developed to a better quality under the influence of Huygens (1629-1695) [28]. In the later part of the 17th century the magnification of telescopes reached 400x and in addition some other improvements were made. These advancements enabled Cassini to discover the first division in Saturn's ring that attests to an effective resolution of almost 1 arc second [29]. During this period Picard (1620-1682) and de la Hire (1640-1718) from the Paris observatory started solar monitoring of unprecedented diligence and regularity under the auspices of a programme directed at the measurement of the Sun's diameter. De la Hire, succeeding Picard after the latter's death in 1682, made observations of solar disc on almost every clear day, on an average of 200 days per year. Picard and de la Hire used telescopes with a focal length of 16 feet (about 5 m) as well as smaller apparatuses with focal lengths from 32 inches to 5 feet [30]. The apertures of these telescopes were 4.0-7.5 cm, *i.e.* their "Raleigh" resolution reached 2-3 arc seconds. According to [30] this was enough to observe sufficiently contrasted spots (pores) with a size of up to 1 arc second. Thus the data of the Paris observatory from 1660-1719 was a valuable source of information on sunspot numbers, which made it possible to trace the increase of the Sun's activity after the Maunder minimum. On 27 December, 1705 Gray in England made the first observation of the white-light flare [31]. In general, the period covering the 17th and beginning of the 18th century was a time of active solar investigation. However, during this period the quality of the observations was far from perfect. For example, on November 4, 1705 de la Hire did not notice the sunspot group observed by four astronomers (Derham, Plantade, Manfredi, Lalande) from three other countries (see [31]). Simultaneous observations by Plantade and Lalande testify that this group lasted for at least 10 days. Thus, according to the rule of Gnevyshev-Waldmeier (formula 1.3), the area of the given group was more than 100 M.S.H. This shows that even in the Paris observatory - the best in that time - observers may have ignored quite large sunspot groups.

After the death of de la Hire (1718) and Cassini (1722) solar astronomy suffered a long period of decline - large telescopes were hardly ever used and general interest towards spots on the Sun declined [32]. As a result, the number of sunspot observations diminished and the observation quality decreased in the middle of the 18th century. Sunspot observations during the Venus transit on 3 June, 1769 clearly illustrate this deterioration. The phenomenon motivated many astronomers to make many associated observations of the solar disc (Table **2.2**). However, the results were rather discrepant: Wright in Canada saw 10 sunspot groups on June 3, 1769, while Darquier in Paris saw only 1 group [33]. As a consequence of the decline of solar astronomy the important discovery of Horrebow, who revealed a regular variability in sunspot numbers in the 1770s, was forgotten [34, 35]. At the end of the 18th and beginning of the 19th century common knowledge about sunspots became so inexact that discussions about the lost solar cycle still continued [36].

Not until in the first part of the 19th century did solar astronomy again start to develop intensively. Schwabe (1782-1875) began regular (200-270 days per year) observations of sunspots in 1826. Rudolf Wolf (1819-1893) started continuous and thorough monitoring of solar activity in 1848. Carrington, Weber, Spörer, de la Rue and many other astronomers followed him throughout the next 20 years. The quality of observation equipment also improved substantially. Joseph Fraunhofer (1787-1826) developed and improved telescope manufacturing technology considerably. He (a) increased the quality of optical glass appreciably

Table 2.2. Observational activity of European astronomers in the 17th-19th centuries (data were taken from Hoyt and Schatten [31, 37].

Observer	Time of Observation, Years	Number of Days with Solar Observations	Relation of the Observation Days with the Total Number of Days
Kirch	1678-1710	481	0.04
Eimmart	1677-1702	2325	0.25
Flamsted	1676-1714	1500	0.10
de la Hire	1682-1718	7170	0.55
Cassini	1700-1709	214	0.07
Plantade	1704-1726	423	0.05
Staudacher	1749-1799	1143	0.06
Horrebow	1761-1776	1532	0.27
Schwabe	1826-1867	11945	0.80
Schmidt	1841-1883	6970	0.45
Shea	1847-1866	5538	0.80
Wolf	1848-1893	10026	0.61
Carrington	1853-1860	1215	0.48
Weber	1859-1883	6983	0.80
Spörer	1861-1893	6283	0.60
de La Rue	1864-1866	451	0.50

due to a new melting technology; (b) carried out a new precise procedure for checking the shape and concentricity of polished lenses; (c) developed a polishing machine so that the lens grinding process became less reliant from the workers' experience, and (d) diminished chromatic aberration by combining lenses made of two types of glass (crown glass and flint glass) with different dispersion. Moreover, the Fraunhofer telescopes had a clock-driven equatorial mounting and were equipped with precise ocular micrometers. Fraunhofer also founded a company, which handled telescope delivery to the largest European observatories. By the mid-19th century Fraunhofer refractors became standard instruments among European astronomers. R. Wolf used the telescope with an aperture of 8 cm, focal length of 110 cm and magnification of 64x [38].

Some fundamental discoveries were made as a result of the qualitatively new level of solar research in the middle of the 19th century. Schwabe reported in 1844 that sunspot activity has a regular variation with a period of ca 10 years. The revealed cycle now is called the quasi 11-year *cycle of Schwabe*. In the 1850-60s,

Spörer (1822-1895) and Carrington (1826-1875) noticed the latitudinal drift of sunspots in the course of the Schwabe cycle (the *law of Spörer*). In 1863 Carrington discovered the differential rotation of the Sun.

In 1858 de la Rue began regular photographic observations of the Sun using the Kew photoheliograph. In 1874 these observations were continued in the Royal Greenwich observatory and an active era of exploration of the physical parameters of solar activity had started [39]. In June 1908 G. Hale first measured magnetic fields in sunspots. This can be considered as the birth of modern astrophysics [17]. Since then the magnetic field of the Sun has become reliably established as the origin of the sunspots. The invention of the photoelectric magnetograph in the middle of the 20th century enabled a thorough research on magnetic fields in the Sun's atmosphere. Modern solar telescopes have apertures of 50-100 cm, a focal length of up to several tens of meters [23], and a resolution of about 0.3 arc second. The satellite era of solar research began in the 1970s. During the last 30-40 years many parameters of solar activity including total solar irradiance, fluxes of energetic particles, X-rays, γ-rays, and UV radiation have been intensively explored by means of spacecraft equipment.

2.3. General Indices of Solar Activity and their Statistical Rules

Modern heliophysics uses statistical (synthetic) as well as physical indices of solar activity [34]. Statistical indexes are derived from data of direct instrumental observations of the Sun using different mathematical algorithms. Usually they have no physical meaning. Sunspot number is the most widespread statistical solar index. On the other hand physical indices reflect the real physical processes linked to solar activity and quantify their measurable manifestations (radio flux, UV flux).

2.3.1. Statistical Solar Indices

Rudolf Wolf was a well-known astronomer from the Zürich observatory and he introduced the relative sunspot number in the middle of the 19th century. The sunspot index initiated by him in the middle of 19th century is called the Zürich or Wolf sunspot number and it is still broadly used as a statistical measure of the Sun's activity. Since 1980 the index is provided and maintained by the Sunspot Index Data Center in Brussels. The Wolf number R_Z is defined as:

$$R_Z = k(10G + F),\qquad(2.5)$$

where G is the number of identified sunspot groups, F is the total number of individual sunspots, and the correction factor k accounts for observational techniques and instrumentation at individual observatories. The Wolf number spans the time interval AD 1700-2010; up to 1996 it was the longest continuous measure of solar activity available for researchers (Fig. **2.1A**).

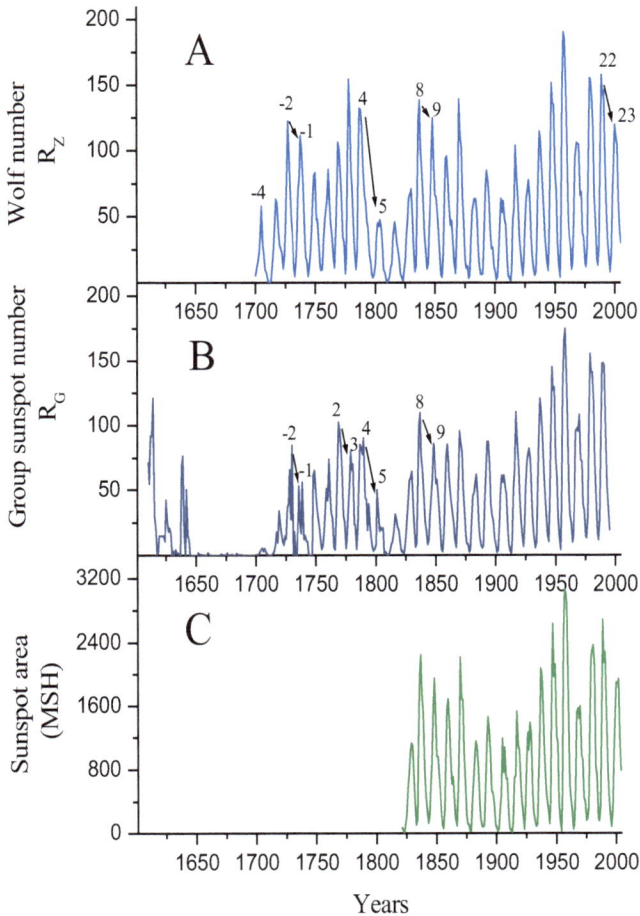

Fig. (2.1). (**A**) Wolf number; (**B**) group sunspot number; (**C**) sunspot area. All the data are yearly averages. Arrows - violation of the Gnevyshev-Ohl (1) rule.

According to the standard (Zürich) cycle numbering the solar cycle 1744-1755 has number 0. The reliability of different parts of the Wolf series is rather different. Eddy [29] graded it into the following periods: poor in 1700-1748, questionable in 1749-1817, good in 1818-1847 and reliable after 1848. The poor quality of Wolf number series in the 18th - beginning of the 19th century is related to scarce and limited observations.

Lately Hoyt and Schatten [37] performed a thorough and extensive archive search and considerably increased the amount of available information. They have introduced a new index of the Sun's activity called the group sunspot numbers R_G. The group sunspot number is defined as:

$$R_G = \frac{12.08}{N} \sum_{i=1}^{N} k_i G_i \quad , \tag{2.6}$$

where N is the number of observers, k_i is the correction factor for observer i, and G_i is the number of sunspot groups identified by observer i, and 12.08 is a normalization factor scaling R_G to R_Z values for the period of 1874-1976, when the number and characteristics of sunspot groups is provided by Greenwich observatory daily reports on. Systematic uncertainties of the annual mean R_G values are estimated to be ca 10% before 1640, less than 5% in 1640-1728 and 1800-1849, 15-20% in 1728-1799, and about 1% since 1849 [37]. Values of R_G are available for AD 1610-1995 (Fig. **2.1B**). The group sunspot number and the Wolf number are nearly identical after the 1870s [37]. However, throughout the major part of the 18th century the difference between R_G and R_Z is appreciable most likely due to the decline of solar astronomy described earlier (Fig. **2.2**).

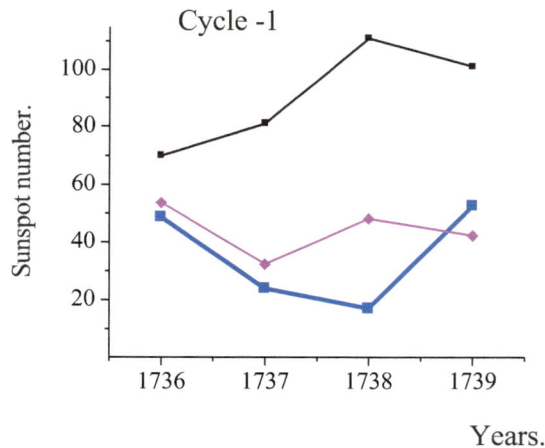

Fig. (2.2). Black line - Wolf number; magenta line - group sunspot number after Hoyt and Schatten [37]; blue line - group sunspot number after Vaquero *et al.* [40]. All the data are yearly averages.

The Wolf number is plotted together with the group sunspot number reconstructed by Hoyt and Schatten [37] and by Vaquero *et al*. [40] who used additional sources of information not available for Hoyt and Schatten. It is evident (see Fig. **2.1A**, **B**) that in 1738 the divergence between different estimations of R_G, performed by Hoyt and Schatten [37] and Vaquero *et al*. [40], is substantially larger than the 20% claimed by Hoyt and Schatten [37]. It should be noted that according to Ogurtsov [41] the precision of determination of yearly values of R_G in the 18th century is less than reported by Hoyt and Schatten [37]. Actual uncertainty is probably close to 30% of the mean R_G value and sometimes can reach a factor of 2 [41].

2.3.2. Physical Solar Indices

Sunspot area measurements for 1874-1976 originate from the Greenwich observatory. The Greenwich series is constructed from day-to-day photographic pictures of the Sun. This series was significantly expanded in [42], who used additional observational data obtained by Spörer, Schwabe, and de la Rue before 1874 and Soviet and Russian astronomers after 1976. They constructed sunspot area series (both monthly mean and yearly average) covering the time interval of 1821-2004 (see Fig. **2.1C**).

This index can be considered as physical, because sunspot area is connected to the sunspot magnetic flux emerging at sunspots. According to Nagovitsyn [39] this relationship is described by the formula:

$$\Phi_\Sigma(t) \approx 2.49 \times 10^{19} A(t), \tag{2.7}$$

where $\Phi_\Sigma(t)$ is magnetic flux in Mx and $A(t)$ is sunspot area in MSH.

Index F10.7 describes the flux of radio emission of the Sun at the wavelength of 10.7 cm (frequency of 2800 MHz). The index is determined in solar flux units ($1\,sfu = 10^{-22}$ W m^{-2} Hz^{-1}). The continuous solar F10.7 record has been available since the year of 1947 (Fig. **2.3B**). F10.7 reflects the ionization of the upper solar atmosphere and correlates closely with sunspot number.

Mg II core-to-wing ratio - The *MgII index* represents the flux of the chromosphere's UV radiation at wavelength about 280 nm [43]. In the core of the Mg II Fraunhofer line, the chromospheric emissions are observable. The Mg II index (Fig. **2.3C**) is obtained by calculating the ratio between the chromospheric

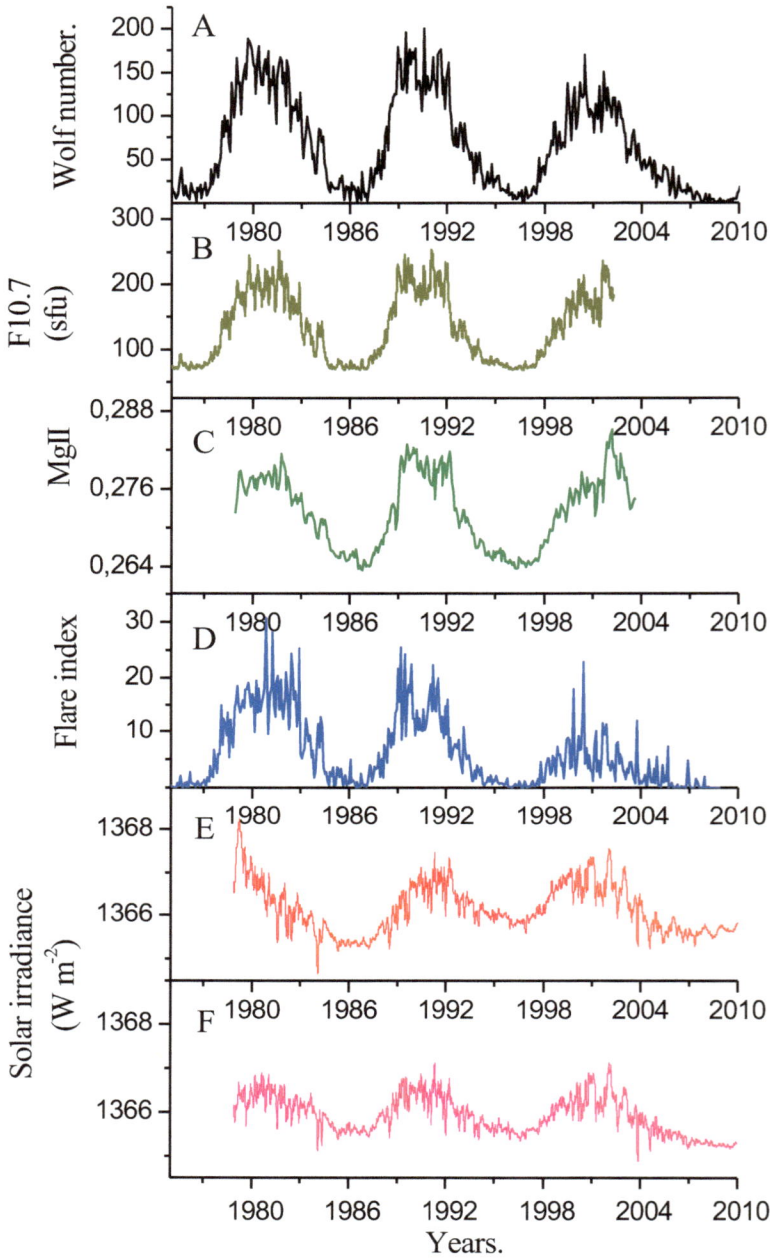

Fig. (2.3). (**A**) Wolf number; (**B**) F10.7; (**C**) MgII index; (**D**) flare index of Klezcek; (**E**) ACRIM series; and (**F**) PMOD series. All the plotted records are monthly mean data.

emission and the solar continuum. It has been measured since 1978 in a series of successive satellite missions. Other UV indices, measured during 2003-2008 are described in [44].

The flare index introduced by Kleczek [44] quantifies daily flare activity by rough estimation of the total energy emitted by the flares. This index is available since 1936 (Fig. **2.3D**).

The *total solar irradiance* or luminosity or TSI (often referred to as the solar constant) is one of the most important physical indexes of solar activity. It is the wavelength-integrated intensity of solar electromagnetic radiation. Solar constant has been measured since the end of 1970s by satellite radiometers. Two composite TSI records - ACRIM series [46] and PMOD series [47] - were calculated from the original satellite data (Fig. **2.3E**, **F**). A close correlation between all the physical indices and sunspot number (see Fig. **2.3**) proves that the Wolf number is a rather precise indicator of the general solar activity changes. Data for Fig. (**2.3**) is from: ftp://ftp.ngdc.noaa.gov/STP/SOLAR_DATA/,http://www.acrim.com/,http://www.pmodwrc.ch/.

2.3.3. General Cyclic Features and Statistical Rules of Solar Activity

The main cyclic features manifested in instrumental solar data are: (a) quasi 11-year cycle; (b) quasi 22-year cycle; and (c) century-type (secular) cycle. The quasi 11-year cycle of Schwabe is the main statistical feature of solar activity. The cycle varies in duration in the range of 8 to 14 years. Eleven-year periodicity is reflected in all solar indices. An idealized form of a quasi 11-year sunspot cycle in sunspot numbers can be described by a formula suggested by Stewart and Panofski [48]:

$$W(\Delta t) = W_0 \, t^a \exp(-b\Delta t), \tag{2.8}$$

where W_0, a, and b are constants for each individual cycle. In 1914 Hale discovered that the polarity of sunspot magnetic fields changes when the next 11-year cycle starts. Hale published the results in 1925 together with Nicholson and the phenomenon is later known as the *Hale-Nicholson law*. Thus the approximate 22-year cycle (the *cycle of Hale*) in which the magnetic polarity of sunspot magnetic field reverses and then returns to its original form is the background for the 11-year Schwabe cycle.

A century-scale cycle in sunspot activity was revealed by Gleissberg [49,50] and is known as the *Gleissberg cycle*. Formerly this cycle was considered as a variation of the 80-90 years period. More recent studies indicate that the Gleissberg cycle has a varying timescale of ca 55-130 years [51, 52] with a bi-modal frequency structure, *i.e.* it consists of 55-80-year and 90-140-year periodicities. Longer solar cycles cannot be studied using instrumental solar observations, but their research requires the use of palaeoastrophysical data.

The Wolf number and sunspot area have a similar distribution pattern (Fig. **2.4**).

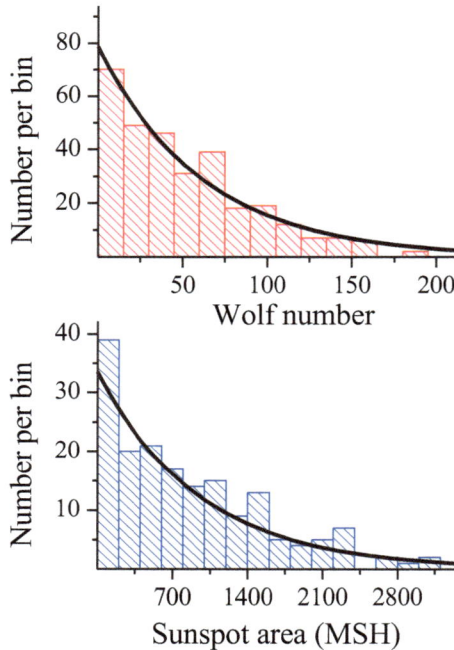

Fig. (2.4). Distribution: (**A**) of yearly averaged Wolf number (1700-2005); (**B**) of yearly averaged sunspot area (1821-1994).

The distributions are well delineated by the formulae of [53]:

$$n(R_z) = 78.66 \ \exp(-0.0163 \ R_z)$$
$$n(A) = 33.48 \ \exp(-0.00105 \ A)$$

(2.9)

where R_Z is the Wolf number, and A is the sunspot area in MSH.

Other important statistical principles of sunspot activity are: (a) Gnevyshev-Ohl rule (even-odd effect); (b) Waldmeier rule; and (c) the amplitude-period effect [54]. The Gnevyshev-Ohl rule [35, 55] is known in several formulations of which the best known ones are:

1)   The amplitude (maximum value) of the odd-numbered cycle with number $2N+1$ is larger than the amplitude of the preceding even-numbered cycle $2N$.

2)   The sum of sunspot numbers integrated over the even-numbered cycle $2N$ has a good correlation with that of the odd-numbered cycle $2N+1$, while the corresponding correlation between cycles $2N$-1 and $2N$ is weak. This is the original formulation of Gnevyshev and Ohl [55].

The Waldmeier rule [56] links the ascending phase of a cycle to the maximum value of sunspot number. It has also two formulations:

1)   The amplitudes of cycles correlate negatively with the lengths of their ascending phases τ (periods between the peak in the sunspot number of a cycle and the time of its starting point). For 23 cycles in the Wolf number this rule has a form [57]:

$$\tau = (45 \pm 12) \ R_{Z,\max}^{-1/2}, \qquad\qquad (2.10)$$

where $R_{Z,\max}$ is the maximum value of the Wolf number over the cycle (cycle amplitude) and τ is the duration of the ascending phase (rise time) in years.

2)   The amplitudes of cycles correlate positively with their rise rates [58].

The amplitude-period effect has two manifestations:

1)   Anticorrelation between the amplitude of a solar cycle N and the length of the previous cycle N-1 as measured from minima to minima. This link has been reported by Chernosky [59], Wilson *et al.* [60].

2)   Anticorrelation between the length of a solar cycle and its amplitude. This relation was mentioned in the works of Dicke [61] and Friis-Christensen and Lassen [62].

All the statistically significant features of solar activity were established empirically. Ogurtsov and Jungner [63] showed that for both the Wolf number and GSSN all the statistical effects are weaker over the time interval AD 1700-1855 than over the time interval AD 1856-1996. This is most likely a result of the uncertainty and less reliability of R_Z and R_G data during the earlier period AD 1700-1855.

2.4. Global Extremes of Solar Activity

Global extremes - extended periods of rise or fall of mean activity level - are another important feature of solar activity. The bulk of these global extremes have been revealed by means of a palaeoastrophysical approach. The Maunder minimum is the only such period which has been established reliably using the data of observations. The prolonged sunspot minimum during the period roughly spanning from 1645 to 1715 was recognized by E. Maunder [64]. Maunder studied the works of G. Spörer, who by examining old records from the different archives of observatories concluded that over the second half of the 17th century sunspots were extremely rare. Later Maunder restricted this period to 1645-1715 [65]. These works were long ignored until 1976 when the concept of the Maunder minimum was reliably confirmed by J. Eddy. The sunspot number (Fig. **2.5**) is determined by means of the data of telescopic observation [37, 66].

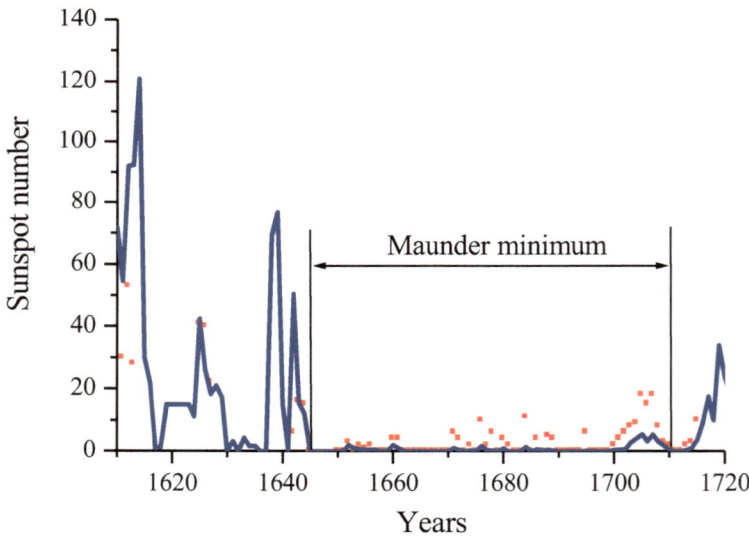

Fig. (2.5). Sunspot number throughout the 17th and 18th centuries. Blue line - data of Hoyt and Schatten [37], red squares - data of Eddy [66].

Even though sunspot observations were rare and sparse throughout the whole of the 17th century they are representative enough to reliably establish close to zero level of solar activity over AD 1645-1715 [34]. Other global solar extremes - Spörer minimum (~1450-1550), Wolf minimum (~1270-1340), Medieval (11th-12th centuries), and Late Medieval (end of the 14th century) maxima *etc.* were revealed by means of the analysis of different solar proxies (see Section 3.7.1).

2.5. Space Weather and Geomagnetic Activity

The Sun influences the Earth and its terrestrial environment in different ways with continuous electromagnetic radiation, magnetic fields, and fluxes of plasma and particles. Solar luminosity is a main source of energy for climatic processes. Solar wind, auroral radiation and energetic particles generated in solar flares encounter the Earth's environment (magnetosphere, ionosphere, and thermosphere) and form the space weather.

Solar wind is a continuous flow of charged particles streaming from the Sun's corona. It consists mainly of protons and electrons. The number of helium ions is 2-4% of the proton number. The mean velocity of solar wind at the Earth's orbit is ca 400 km/s and thus its travel time from the Sun to the Earth is of the order of 80 hours [67]. Solar wind carries the Sun's magnetic field out into the space. At the Earth's orbit the interplanetary magnetic field (IMF) has a mean magnitude of about 5 nT (5×10^{-5} G). According to Svalgaard and Cliver [68] it changes with the sunspot activity as:

$$B = a + bR_Z^{1/2},$$ (2.11)

where B is in nT, $a = 4.62 \pm 0.16$, $b = 0.272 \pm 0.015$, and R_Z is the Wolf number. Monthly values of the solar wind velocity and IMF are measured by various spacecrafts near the Earth's orbit (Fig. **2.6D**, **E**). B_z - the southward (negative) component of IMF - controls the energy exchange between solar wind and the magnetosphere-ionosphere system. Solar wind has both slow and fast components. High-speed flows emanating from *coronal holes* - low-density parts of the corona with open magnetic field lines - dominate in solar wind during the declining and minimum phases of the solar cycle. This open structure allows charged particles to escape from the Sun. A denser low-speed component is associated with the equatorial coronal streamer belt. At solar maximum and during the ascending phase of solar activity cycle the average speed of solar wind decreases and the flow is substantially disturbed by *coronal mass ejections* (CME) - the huge eruptions of coronal plasma from closed magnetic field domains that expel gigantic (up to 10^{10}-10^{11} cm) clouds of ionized gas into interplanetary space. The mass of CME is 10^{15}-10^{16} g and energy up to 10^{31} erg [68, 69]. The maximum velocity of high-speed solar wind fluxes is proportional to the coronal hole area and can reach 2,000 km/s. There are two main types of solar-induced interplanetary ejecta: magnetic clouds and complex ejecta. Magnetic clouds have a flux rope geometry of magnetic field. Their magnetic field structure was described by numerous models (*e.g.* [70-72]). Complex ejecta have disordered (no flux rope structure) and weaker magnetic fields [73]. The frequency of CME occurrence tends to follow the solar cycle. According to Gopalswamy *et al.* [74] the rate of occurrence of the large (the apparent angular width >30°) CME is:

$$V(CME > 30^0) = 0.02 \times W + 0.9,$$
(2.12)

where V is in day^{-1} and W is the Wolf number.

Table 2.3. **Main characteristics of the solar-induced corpuscular radiation in near-Earth space.**

	Particles	Velocity (km/s)	Density of Particles (cm^{-3})	Density of the Energy Flux (erg× cm^{-2}×s^{-1})	Time of Flight to the Earth	Energy of Particle	Depth of Penetration
Solar wind	protons	200-400	7-15	≈1.5	3-4 days	0.5-1 keV	magnetosphere
	electrons	200-400	7-15	<0.1	3-4 days	≈ 0.5 eV	
Fast solar wind	protons	400-800	3-4	≈1.4	1-2 days	up to 3-5 keV	magnetosphere
	electrons	400-800	3-4	<0.1	1-2 days	≈ 1.5 eV	
Auroral radiation	protons	up to 3000	<10^{-2}	<0.1	<1 day	10-100 keV	80-100 km
	electrons	up to 5×10^4	up to 100	up to 5000	few hours	1-30 keV	
Subrelativistic PCA particles	protons	up to 10^5	~10^{-6}	0.2-5.0	δ1 hour	10-100 MeV	35-60 km
SCR	protons	≤3×10^5	~10^{-8}	~0.2	~10 minutes	up to 10-20 GeV	up to the ground

Disturbances in the flux of the magnetized solar plasma hitting the Earth shake the magnetosphere and cause fluctuations in geomagnetic activity, which are manifested as geomagnetic storms, sub-storms and aurorae. *Geomagnetic storms* are the most important disturbances of the magnetosphere related to exclusive solar wind events. They happen when the IMF turns southward and keeps this direction for an appreciable time interval. A geomagnetic storm consists of three phases: (a) initial phase; (b) main phase; and (c) recovery phase. The value of the D_{st} index is a good indicator of a geomagnetic storm. D_{st}, or storm-time variation index is the hourly average of excursion of the H (horizontal) component of the geomagnetic field compared with an undisturbed time. It is determined using the data from four low-latitude geomagnetic stations (three in the northern hemisphere and one in the southern hemisphere). During the initial phase of the geomagnetic storm D_{st} often increases for many hours. The main phase of the storm takes place over a period of about one day. During this stage charged particles in the circumterrestrial plasma sheet are accelerated and injected into the inner layers of magnetosphere, creating the storm-time ring current. This causes a decrease of the H component by 50-150 nT during weak storms, which occur about once in a month [2], and by 300-400 nT or more during very large storms, which occur a few times over a solar cycle. The strongest geomagnetic storm was a result of a white light solar flare on 1-3 September, 1859 (the Carrington Event). During September 2-3 the D_{st} values reached 1760 nT in Bombay ($\Phi = 18.9°N$) [75], more than 1000 nT in Ekaterinburg ($\Phi=49.5°N$) and more than 700 nT in St. Petersburg ($\Phi = 60.0°N$) [75]. The recovery phase lasts for several days. Geomagnetic storms are accompanied by an injection of a large number of particles into the outer radiation belt. *Sub-storms* take place over a period of a few hours and are observable primarily at high geomagnetic latitudes. They occur several times per day and cause disturbances in the auroral zone in the range of 200-2000 nT. Thus, the variation of the geomagnetic field during storms and sub-storms does not exceed a few percent of its mean value (3×10^4 nT). Some indices are used for a description of geomagnetic activity besides the D_{st} index. The K-index is a quasi-logarithmic local index derived from the maximum fluctuations of the horizontal components of the geomagnetic field observed by a magneto-meter through a three-hour interval. It was first introduced by J. Bartels in 1938 and it has an integer value in the range 0-9, with K_p= 0-1 being calm, K_p=5 is a minor geomagnetic storm and $K_p \geq 6$ or indicating a major storm. The planetary 3-hour-range index K_p is the mean standardized K-index from 13 mid-latitude magnetometer stations (11 in the northern hemisphere and two in the southern hemisphere). The scale is 0 to 9 expressed in thirds of a unit. In contrast to the D_{st} index, which is highly sensitive to geomagnetic storms, the K_p index also reflects sub-storm activity. The relationship between K_p value and solar wind parameters can be described by an approximate formula of Goncharova and Maltsev [76]:

$$K_p = \frac{n_p + 70}{8000} V - 2 \cdot (1 - \frac{B_s}{8}), \qquad (2.13)$$

where $B_s(nT) = \begin{bmatrix} B_z, B_z < 0 \\ 0, B_z > 0 \end{bmatrix}$, V is the velocity of solar wind in km/s, and n_p is the

concentration of protons in solar wind in cm^{-3}. The Aa index (in units of 1 nT) is the simplest three-hourly global geomagnetic activity index, which is calculated from the K_p indices from two approximately antipodal observatories. At the present time these are the Canberra observatory in Australia and the Hartland observatory in the UK. Both storms and sub-storms inject many auroral particles into the upper atmosphere. *Auroral radiation* is a flux of charged particles, which causes auroral phenomena (aurora borealis, aurora australis, and magnetic disturbances) in the high-latitude ionosphere. *Aurorae* or polar lights are the large luminous areas, which emerge in the sky and form arcs, rays, bands and spots. In general they appear at the altitudes of 80-150 km. During a quiet geomagnetic time ($K_p = 0$-1) polar lights can be seen only close to the auroral oval ($\sim67°$) while during strong storms ($K_p = 8$-9) they can appear at latitudes 50° or less. The most southern point at which aurora borealis has ever been observed (15 September, 1909) is Singapore. Arorae are caused by intrusions of the auroral particles - protons with energy of up to 50-100 keV and electrons with energy of up to several tens of keV - which ionize and excite molecules of atmospheric gases:

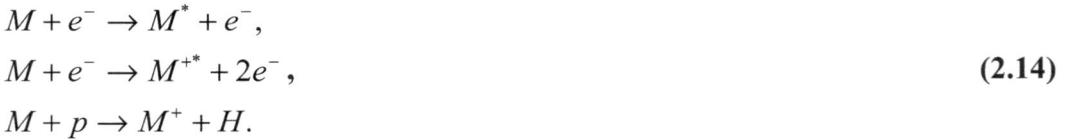

$$M + e^- \rightarrow M^* + e^-,$$
$$M + e^- \rightarrow M^{+*} + 2e^-,$$
$$M + p \rightarrow M^+ + H.$$

(2.14)

where M consists of N, N_2, O, and O_2. The excitation causes electromagnetic emission and luminescence. Electrons play the primary role in the auroral processes. Auroral particles originate from particles of high-speed solar wind and flares trapped in the magnetosphere and accelerated up to auroral energies during reconnections of magnetic lines in the magnetosphere's tail. Such reconnections occur usually during geomagnetic sub-storms and storms, when fluxes of solar wind increase due to disturbances on the Sun. Accelerated particles move down the magnetic field lines of the Earth enter the upper atmosphere and collide with its atoms and molecules of the upper atmosphere. A minor part (about 1%) of polar light can be produced directly by solar flare particles without additional acceleration in the magnetosphere. Atomic oxygen is responsible for two main auroral colours: the most frequent green-yellow light (558 nm), and the deep red light (630 nm) which is seen less frequently. Purplish-violet light can be observed over high (>1000 km) altitudes due to the luminescence of the ion N^+ [2]. The frequency of auroras is linked to the level of solar activity. Throughout the years of 1780-1829 the number of northern lights observed over middle Europe per year can be approximately described with a formula of Schröder *et al.* [77]:

$$N_{Aur}(R_Z) = (21 \pm 1) \cdot \frac{R_Z}{100} \cdot (1 + \frac{R_Z}{100}),$$
(2.15)

where N_{Aur} = year^{-1} and R_Z is the Wolf number. Aurora borealis is a magnificent natural phenomenon which undoubtedly has attracted people and may thus have been easily observed. That is why first aurora records date back more than 2 000 years.

Powerful solar flares are the main sources of the subrelativistic and relativistic solar energetic particles (SEP), the energies of which can reach several tens of MeV and sometimes extend into the GeV range. An acceleration of particles by shock waves associated with CMEs can take some part in the SEP generation. More than 90% of solar energetic particles are protons. They can hit the Earth within 15 minutes to 2 hours after the solar flare. The energy spectrum of SEP can be described with the formula:

$$\frac{dN(R)}{dR} \approx \exp(-R/R_0),$$
(2.16)

where $N(R)$ is in cm^{-2} s^{-1} MV^{-1} sr^{-1}, $R = \frac{pc}{Q}$ is the magnetic rigidity of particle in MV, p is the momentum, Q is the charge, and c is light velocity. R_0 is the characteristic rigidity, which is different for different flares. Typically R_0 lays in the range 50-100 MV, but for a very powerful flare on February 23, 1956 its value reached 325 MV. A *Solar proton event* (SPE) is the sharp increase of emission of solar energetic particles in which proton flux with energy $E_p > 10$ MeV is greater or equal to 10 proton flux units (pfu) or 10 particles cm^{-2} s^{-1} sr^{-1} [78]. SPE lasts from a few hours to several days.

To the first order the Earth's magnetic field can be described as a dipole with the moment 8×10^{22} A m^2. The geomagnetic field prevents charged particles from hitting the Earth and thus works as a shield. The *cutoff rigidity* R_c is the minimum magnetic rigidity of a charged particle that can reach a given point at the Earth's surface. This value depends on the geographic coordinates (particularly latitude) of the point (for its approximate estimation, see Appendix 1).

Fluxes of subrelativistic (E_p = 1-100 MeV) protons cause very strong attenuation of radio waves over the polar regions due to the extremely heavy ionization of the D and E layers (60-120 km) of the ionosphere. These events are *polar cap absorptions* (PCA). They can be observed by means of a riometer (relative ionospheric opacity meter), which measures the absorption of the cosmic radio emission at high frequency (20-60 MHz). Relativistic protons the energy of which reaches to 10-20 GeV are called (especially in Russian scientific literature) *solar cosmic rays* (SCR). In extreme cases fluxes of SCR can reach lower atmosphere be detected by ground-based neutron monitors mounted on the ground. As a result *ground level enhancements* (GLE) occur. Solar flares with energy of 10^{-4} W m^{-2} produce a GLE event with a probability of 10% and with energy 10^{-3} W m^{-2} - with a probability of 70% [79]. At the time of the solar flare of February 1956 the counting rate of a neutron monitor located at the South Pole increased by about 50 times [80] and during the flare of January 2005 - more than 50 times [81]. Neutron monitors situated at lower latitudes (geomagnetic cutoff 0.4-0.6 GV) reacted to the event of January 2005 by an increase in counting rate up to 400% [81]. The GLE of October 1989 caused a rise of counting rate of Huancayo monitor (geomagnetic cut-off 13 GV) by more than 10%. Evidence that SCR might contain ultra-relativistic protons of higher energies was obtained recently. For example, during the GLE of October 1989 the muon spectrometer of collaboration L3 detected an increase of muon flux that can be provided by a substantial flux of protons with energy more than 40 GeV [82]. Some proton events were manifested in the Baksan muon detector by the growth of muon flux indicating that primary protons, which originated these muons, have energies of more than 500 GeV [83]. The general features of solar wind, auroral radiation and solar energetic particles are summarized (Table **2.3**) from several studies [85, 86].

Low-rigidity cosmic particles can be trapped in the magnetosphere by the Earth's magnetic field. These particles form radiation or Van Allen belts. The inner radiation belt covers the altitude range 0.06-2.0 radii of the Earth (R_E) above the surface. It contains high concentrations of electrons with energies of 20 keV-1 MeV and energetic protons with energies of 20-800 MeV or more. Presence of ultra-relativistic electrons with energies more than 100 MeV was reported by Galper *et al.* [84]. The outer radiation belt extends from an altitude of about three to ten R_E above the Earth's surface. It contains electrons with energies from 40 keV to several MeV. The total number of particles trapped in radiation belts is 10^{29}-10^{30}. The majority of them have a solar origin.

During geomagnetically disturbed periods high-energy (few hundreds of keV) electrons of the radiation belts can precipitate into the Earth's atmosphere [87]. Such episodes are called *relativistic electron precipitations* (REP) events. The results of measurement of precipitations of energetic (E > 200 keV) electrons carried out by the Lebedev Physical Institute from 1957 till now [87-91] using radiosonde-detectors in the atmosphere are shown in Fig. **2.6F**.

The relationship between solar flares, coronal mass ejections, the generation of magnetic clouds and large (K_p > 5) geomagnetic storms is described in Tables **2.4** and **2.5** (the data were taken from [75, 87-91]).

Table 2.4. A link between the two phenomena: X resulted in Y.

X	Y	Probability
CME	MC/E	0.60-0.80
MC/E	Storm	0.40-0.60
CME	Storm	0.35-0.50
Flare	CME	≅0.10
Flare	Storm	0.30-0.40

Table 2.5. A link between the two phenomena: X is a result of Y.

X	Y	Probability
CME	Flare	≅0.40
Storm	Flare	0.50-0.80
MC/E	CME	0.50-0.80
Storm	MC/E	0.30-0.70
Storm	CME	0.80-1.00

In Tables **2.4** and **2.5** CME means coronal mass ejections and MC/E - magnetic cloud and ejecta geomagnetic storms can be linked to coronal mass ejections [90].

The heliosphere has abundant amounts of galactic cosmic rays whose flux is constant and isotropical. GCR have energies of up to 10^{20} eV. Their energy spectrum has the form:

$$\frac{dN(E)}{dE} \approx E^{-\gamma}, \tag{2.17}$$

where $N(E)$ is in cm^{-2} s^{-1} GeV^{-1} sr^{-1} and $\gamma = 2.6$ (when $E = 10\text{-}10^6$ GeV). GCR consist of a nuclear component (protons - 92%, α-particles - 7%, C, N, O nuclei - 0.5%) and electrons, the number of which is ca 10% of the nuclei number. Protons of GCR with energy 1 GeV, which can penetrate deep into the atmosphere, have a density of ca 10^{-9} cm^{-3} at the Earth's orbit. Their energy flux is about 3×10^{-2} erg cm^{-2} s^{-1} [53]. Galactic cosmic rays are observed with ground-based neutron monitors, which are particularly sensitive to particles with energy of several GeV. A more detailed description of GCR can be found in [92, 93]. Solar activity influences the intensity of low-energy (up to several tens of GeV) GCR. Present-day solar modulation of cosmic rays is produced by three major mechanisms [94]:

(a)   Outward convection of the magnetic field and GCR caused by the solar wind flux.

(b)   Diffusion of GCR caused by their scattering by the irregularities in the IMF. The density of heliospheric magnetic inhomogeneity depends on the Sun's activity level. When solar activity increases the density of these heterogeneities rises. Solar wind velocity also shows some positive correlation with solar activity (see Fig. **2.10**). That is why the diffusion-convection effects cause an 11-year variation of GCR intensity. The flux of cosmic ray particles with energy $E = 0.1\text{-}15$ GeV during solar minimum is twice as large as during the maximum phase [95, 96].

(c)   The drift modulation of GCR caused by changes of the polarity of the solar magnetic field. Positively charged cosmic rays preferentially enter the heliosphere from the direction of solar poles during a time when the solar magnetic field in the northern hemisphere has negative polarity, *i.e.* is directed outwards [2]. As a result, cosmic ray time behaviour has a peaked form during solar cycles with negative polarity and it has a flat-topped form during cycles with positive polarity. Since the solar magnetic field changes its polarity after every 11 years, the cosmic ray flux also follows a 22-year cycle with alternate maxima being peak-typed and plateau-type. The amplitude of the 22-year variation in the integral value of the GCR flux, registered by a neutron monitor, reaches 5% or more [97]. Theoretic calculations performed in [98] show that during the global Maunder minimum the amplitude of drift variation could reach 15% for particles with $E = 0.5\text{-}50$ GeV.

Solar activity can also influence GCR intensity over a short timescale. A decrease of GCR intensity during several days after the large flares on the Sun and intensive solar CME is called a *Forbush decrease* (FD). The amplitude of a Forbush decrease typically is 2-5% for GCR with $E = 500$ MeV. A usual Forbush decrease starts as an abrupt (within minutes) decrease of a cosmic ray flux. The recovery phase begins after a few hours and can last for several days during which the GCR intensity returns to normal. Mishra *et al.* [99], who analyzed 41 FDs during 1996-2006, found that 88% of them were associated with CMEs and 55% with bright solar flares. Some physical parameters related to geomagnetic activity and space weather are shown in Fig. (**2.6**) together with Wolf number. Data were taken from the sites: ftp://ftp.ngdc.noaa.gov/STP/SOLAR_DATA/COSMIC_RAYS/STATIO N_DATA/Climax/docs/climax.tab, http://nssdcftp.gsfc.nasa.gov/spacecraft_data/o mni/omni2.text and Makhmutov *et al.* [88].

Solar electromagnetic and corpuscular radiation that reaches the magnetosphere and atmosphere can also affect the space near the Earth and cause changes in space weather. The depth of penetration of solar radiation into the Earth's atmosphere depends on the radiation type (Fig. **2.7**).

2.6. A Brief History of Solar-Terrestrial Research

In 1716, E. Halley (1650-1742) suggested that aurora is caused by "magnetic effluvia" moving along the Earth's magnetic field lines. This educated guess determined the first step in the Sun-Earth connection science [67]. G. Graham (1675-1751), A. Celcius (1701-1744) and O. Hiorter (1696-1750) observed time coincidence between compass needle fluctuations and a bright auroral display in March 1741 [103]. Thus, the relationship between individual aurorae and geomagnetic disturbances was settled. In 1754 J.-J. Mairan (1678-1771) noted a connection between sunspots and aurorae [103]. J. Ritter (1776-1810) in 1803 established maxima in the occurrence of aurorae at 1720/1723, 1728/1732, 1739/1742, 1751, 1760, 1769/1770, 1779, and 1788 [104]. This was the first indication of an 11-year cycle in aurorae. J. von Lamont (1805-1879) in 1851 noted a possible period of 10.33 years in magnetic declination data. E. Sabine

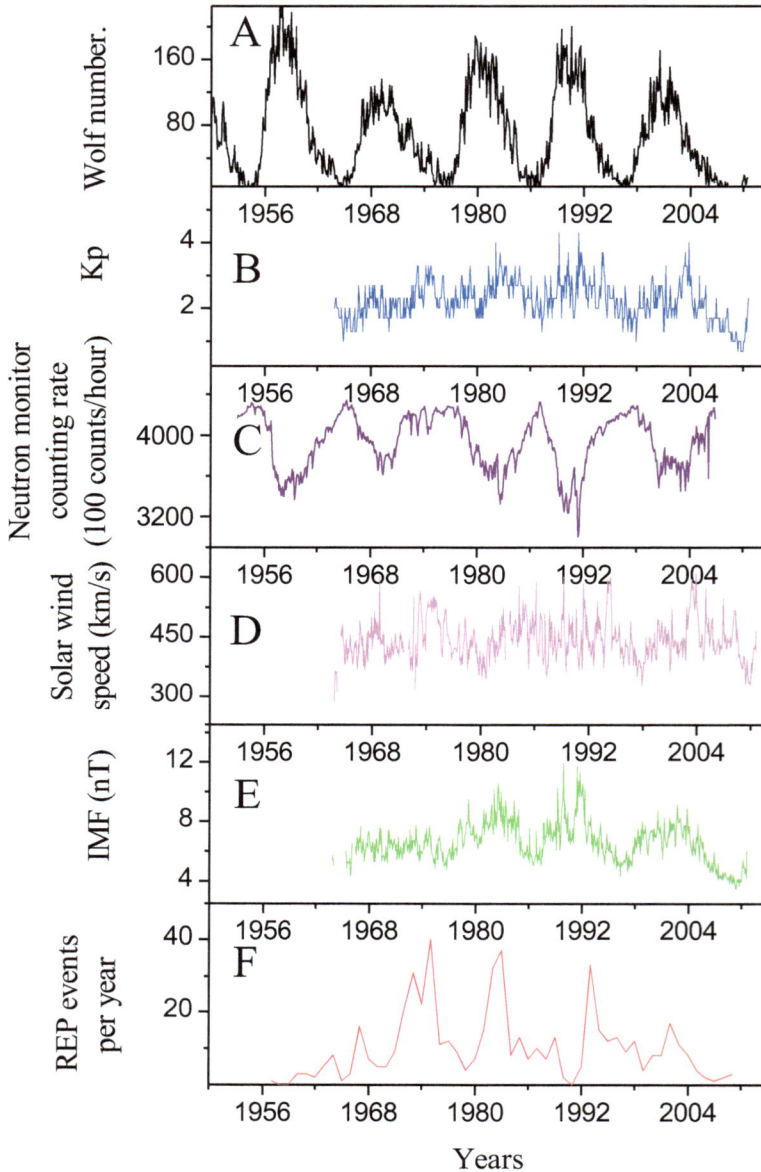

Fig. (2.6). (**A**) Wolf number, (**B**) geomagnetic K_p index; (**C**) counting rate of Climax neutron monitor (39°N, 106°, W, R_c=3 GV); (**D**) - velocity of solar wind measured near the Earth's orbit; (**E**) intensity of interplanetary magnetic field measured near the Earth's orbit; (**F**) yearly number of relativistic electron precipitation events recorded in the stratosphere at Murmansk region (magenta line, geomagnetic cutoff rigidity R_c=0.6 GV). All the plotted records, besides REP, are monthly mean data.

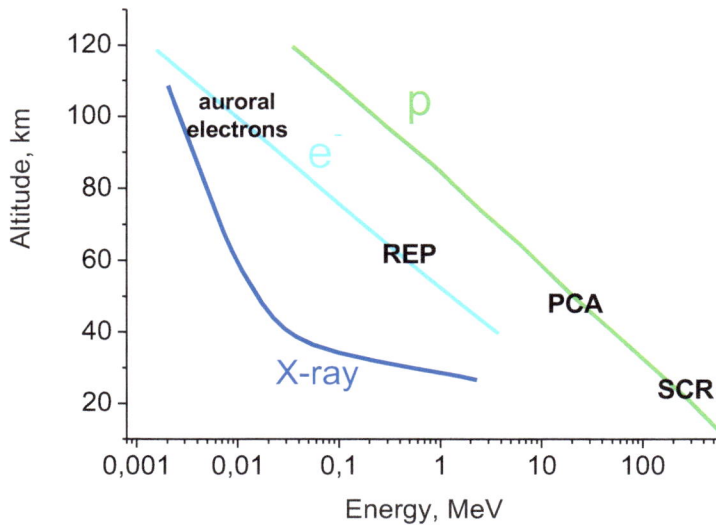

Fig. (2.7). The altitudes in the atmosphere, which could be reached by vertically moving protons, electrons and X-rays. Data were taken from [100, 102].

(1788-1883) also found in 1852 a periodicity of approximately 10 years in the incidence of geomagnetic disturbances in Canada. He related its origin to periodical changes of the Sun. Carrington revealed a relationship between solar flares, bright auroras and geomagnetic disturbances in 1860. Finally, a connection between geomagnetic activity and solar processes was established by the mid-19th century. In the beginning of the 20th century K. Birkeland (1867-1917) suggested that sunspots might be the source of particles that influence terrestrial magnetosphere and thus introduced the conception of plasma flow from the Sun [2]. The satellite epoch of solar-terrestrial research started at the end of the 1950s and brought important findings. Radiation belts were discovered in 1958 due to experiments performed using the satellites Explorer-1 and Explorer-2 by a group of American scientists headed by J. Van Allen. The inner radiation belt was first detected on 7 November 1957 by a device mounted on the second Soviet satellite (the experiment was led by S. Vernov and A. Chudakov). However, Soviet scientists did not interpret their finding correctly [105]. The first direct measurement of solar wind flux outside the magnetosphere was made in 1959 by means of ion traps on board the spacecrafts Luna-2 and Luna-3 [106]. In the early 1970s measurements made by Skylab equipment confirmed the existence of ionized matter clouds caused by coronal mass ejections and already proposed by Chapman and Ferraro in the 1930s [2].

Cato the Elder (234-149 BC) mentioned a relation between the darkening of the solar disc and rye prices [107]. Furthermore, B. Balliani noted that sunspots do work as coolers, and thus in the case of many spots on the Sun one can anticipate lower temperatures on the Earth. A conspicuous relationship between the prices of some weather-dependent harvest of farm products and sunspot numbers was revealed by W. Hershel in 1801. The work of Hershel was a starting point for serious Sun-climate research. G. Wild [108] was probably the first who noted the effect of solar flares and geomagnetic disturbances on the surface temperature. Ney [109] suggested that the flux of galactic cosmic rays is an agent transferring solar influence on climate. Dickinson [110] found that the fluxes of cosmic rays might influence cloudiness. Currently the study of a possible solar-climate relationship is an important branch of solar-terrestrial physics. Extraction of information from numerous solar, climatic and environmental proxy series is in turn a significant part of mathematical stutistics, which is being actively developed at present.

2.7. Statistical Analysis of Solar and Geophysical Time Series

2.7.1. Stationary and Non-Stationary Time Series

With statistical analysis of natural, *e.g.* solar or climatic scalar (one-dimensional) time series, we refer to the use of methods and approaches of mathematical statistics aimed at the extraction of information on the character and dynamics of natural processes which generate a specified signal. Therefore, revealing the components regularly varying in time, the analysis of their time evolution and estimation of their amplitudes, the detection of repeating acyclic structures, as well as the evolution of the character and power of noise deforming a signal, can be considered as the main objectives of such statistical analysis. Statistical analysis of natural (solar, climatic, seismic, acoustic *etc.*) time series is complicated because the majority of them are strongly non-stationary while the common classical methods of mathematical statistics were developed on the assumption of the stationary character of investigated processes.

A stationary process is a random (stochastic) process, the probability density (distribution function) of which does not depend on time. White noise (a set of non-correlated random numbers) is an elementary random process. If this noise $y(t)$ is Gaussian its distribution function has the well-known form:

$$P(y) = \frac{1}{\sigma\sqrt{2\pi}} \exp(-\frac{1}{2}\left(\frac{y - \bar{y}}{\sigma^2}\right)^2)$$ (2.18)

where $\bar{y}$ is the average and σ^2 is the variance of $y(t)$. A non-stationary process has a probability density that changes in time. The general types of non-stationary are:

1)   change of the average of a time series in time

2)   change of the spectral content in time of a time series in time

3)   change of the variance in time

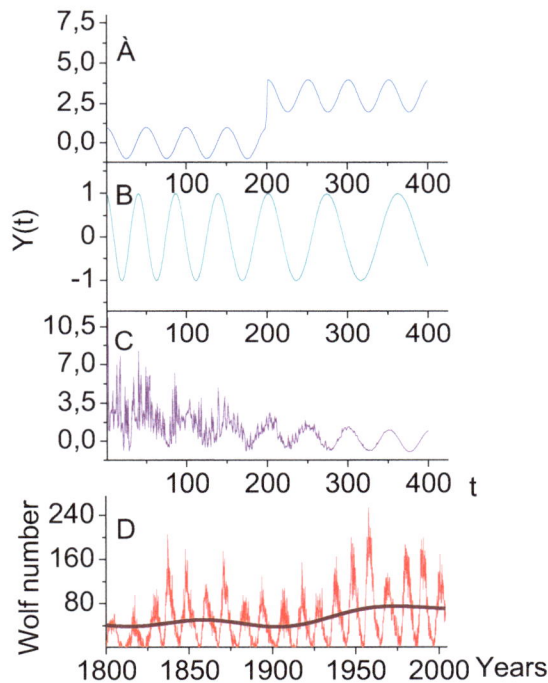

Fig. (2.8). (A) signal $Y(t)$ with mean value changes in time; **(B)** signal $Y(t)$ with spectral content changes in time; **(C)** signal $Y(t)$ with variance changes in time; **(D)** monthly Wolf sunspot number (red curve) and its mean value (brown curve).

All three types of non-stationary (Fig. **2.8**) are usually inherent in natural time series, monthly Wolf numbers as an example (Fig. **2.8D**). This sunspot series shows temporal changes in the average value, in spectral content (the cycle period varies from 8 to 17 years) and in variance (the power of noise in a minimum of a cycle is much less than in a maximum).

2.7.2. Statistical Analysis of Non-Stationary Time Series

Strong non-stationary of the actual solar and geophysical time series complicates the application of classical methods of spectral analysis. For example, Fourier transform, described in a huge body of literature and often used as a representation of a time-domain signal in the frequency domain, is not an effective tool in the case of a quasi-periodic non-stationary process. The normalized Fourier spectrum $P_N(\omega)$ of a discrete time series $Y(t)$, which consists of N points and has a Fourier transform:

$$\hat{Y}_k = \frac{1}{N}\sum_{n=0}^{N-1} Y_n \cdot \exp(-2\pi ikn/N), \; k\text{=0...}N\text{-1}, \tag{2.19}$$

is determined as the square of the Fourier transform divided by the total variance (see *e.g.* Torrence and Compo [111]):

$$P_N(\omega) = \frac{N \cdot \left|\hat{Y}(\omega)\right|^2}{\sigma^2}. \tag{2.20}$$

As a result $\int_0^{\omega_{max}} P_N(\omega)d\omega = 0.5$, where $\omega_{max} = \dfrac{1}{2\cdot\Delta t}$ is called Nyquist frequency and Δt is sampling interval. The Fourier transform decomposes a signal into orthogonal trigonometric basis functions. It can estimate only the average power of the time series at a given frequency, and it can not to trace the evolution of in spectral content through time.

This problem is overcome using a continuous wavelet transform. The wavelet transform differs from the Fourier transform in that the analysed signal is not decomposed into infinite sinusoids but into a number of orthogonal waves, which are called wavelets and are well localized both in the frequency and in the time domains, while Fourier harmonics are localized only in the frequency domain. Because of these properties, the wavelet transform is suitable for analyses of non-stationary time series, including non-periodic non-homogeneities, deterministic chaos and local periodic structures [112]. The wavelet approach, considerably developed recently by [111-115], is based on the examination of the spectrum of $W^2\ (a,\ t)$, the square of coefficients of the wavelet transform of the analyzed time series. These coefficients $W\ (a,\ t)$ are determined by means of the formula:

$$W(a,t) = a^{-1/2} \int\limits_{-\infty}^{+\infty} Y(t')_* \psi^* \left(\frac{t - t'}{a} \right) dt \ , \tag{2.21}$$

where $Y(t)$ is the analysed signal, $\psi(t)$ is the analysing wavelet, a is the scale parameter of the analysed wavelet, which determines the wavelet size, and t' determines the translation of $\psi(t)$, *i.e.* specifies the wavelet localization. The wavelet approach allows extracting reliable information about the spectral structure of non-stationary signals and their time evolution. Its great utility is that it gives both spatial and frequency resolutions. The wavelet analysis is often considered as some kind of 'a mathematical microscope' which allows investigation of the internal structure of complicated non-uniform objects. Statistical analysis of time series is usually performed using wavelet spectra (or energy spectra) - the squares of the coefficients of the wavelet transform $w\ (a,\ t)$. The value $w\ (a,\ t)^2$ is called a local wavelet power spectrum. It is also known as the density of energy of a signal [112]. The value $w\ (a)^2$ - integrated over the time coefficient $w\ (a,\ t)$ - is called a global wavelet power spectrum or the scalogram of a signal. The complex bases of Morlet and real MHAT (Mexican Hat) are suitable for studying non-stationary data sets. The basis of Morlet has a form of:

$$\psi_{Morle}(t) = \pi^{-1/4} \cdot \exp(-t^2) \cdot \exp(-ik_0 t) , \tag{2.22}$$

(k_0 usually is equal to 6.0 to satisfy the admissibility condition) and its Fourier transform is:

$$\hat{\psi}_{Morle}(\omega) = \pi^{-1/4} \cdot H(\omega) \cdot \exp(-(\omega - \omega_0)^2 / 2) \tag{2.23}$$

where $H(\omega)$ is Heaviside function. MHAT-basis and its Fourier transform are described by the formulas:

$$\psi_{mhat}(t) = \exp(-t^2 / 2) \cdot (1 - t^2) , \tag{2.24}$$

$$\hat{\psi}_{mhat}(\omega) = 2\omega \cdot \exp(-\omega^2 / 2). \tag{2.25}$$

The set of scales a_j was chosen according to [106]:

$$a_j = a_0 \cdot 2^{j\vartheta} \text{ is the time scale, } j = 0,1, ..., J; \qquad (2.26)$$

$$J = \vartheta^{-1} \cdot \log_2 (\frac{N_0 \cdot \delta t}{s_0}),$$

where δt is the sampling interval of the series, N is the number of points in the series, δj determines the timescale step, and s_0 determines the minimal resolvable timescale (usually $s_0 = 2 \cdot \delta t$). For a more convenient comparison of different wavelet spectra, these spectra are usually normalized by variances.

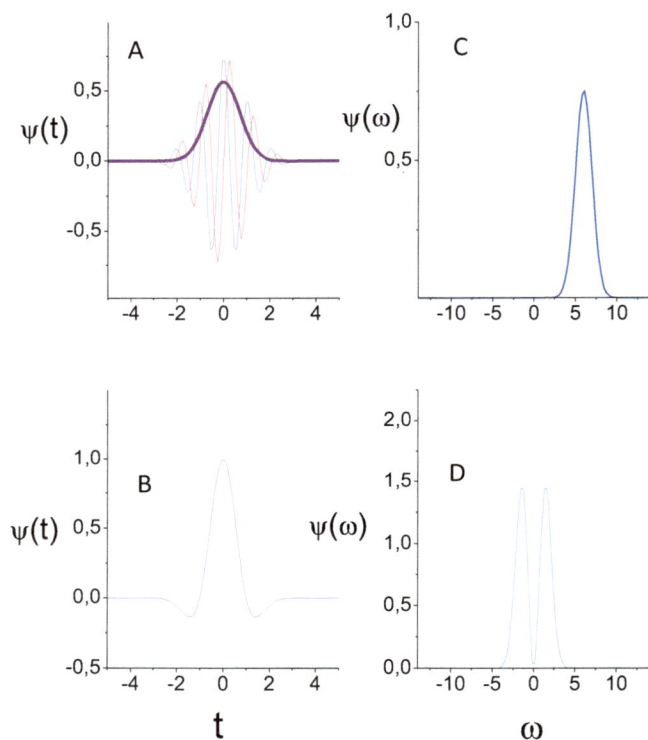

Fig. (2.9). (**A**) Morlet wavelet basis. Violet plot - real part, red plot - imaginary part, bold purple plot - absolute value; (**B**) MHAT wavelet basis; (**C**, **D**) their Fourier transforms.

The MHAT wavelet is narrower than the Morlet wavelet in time space and broader in frequency space (Fig. **2.9**). This means that using the Morlet basis

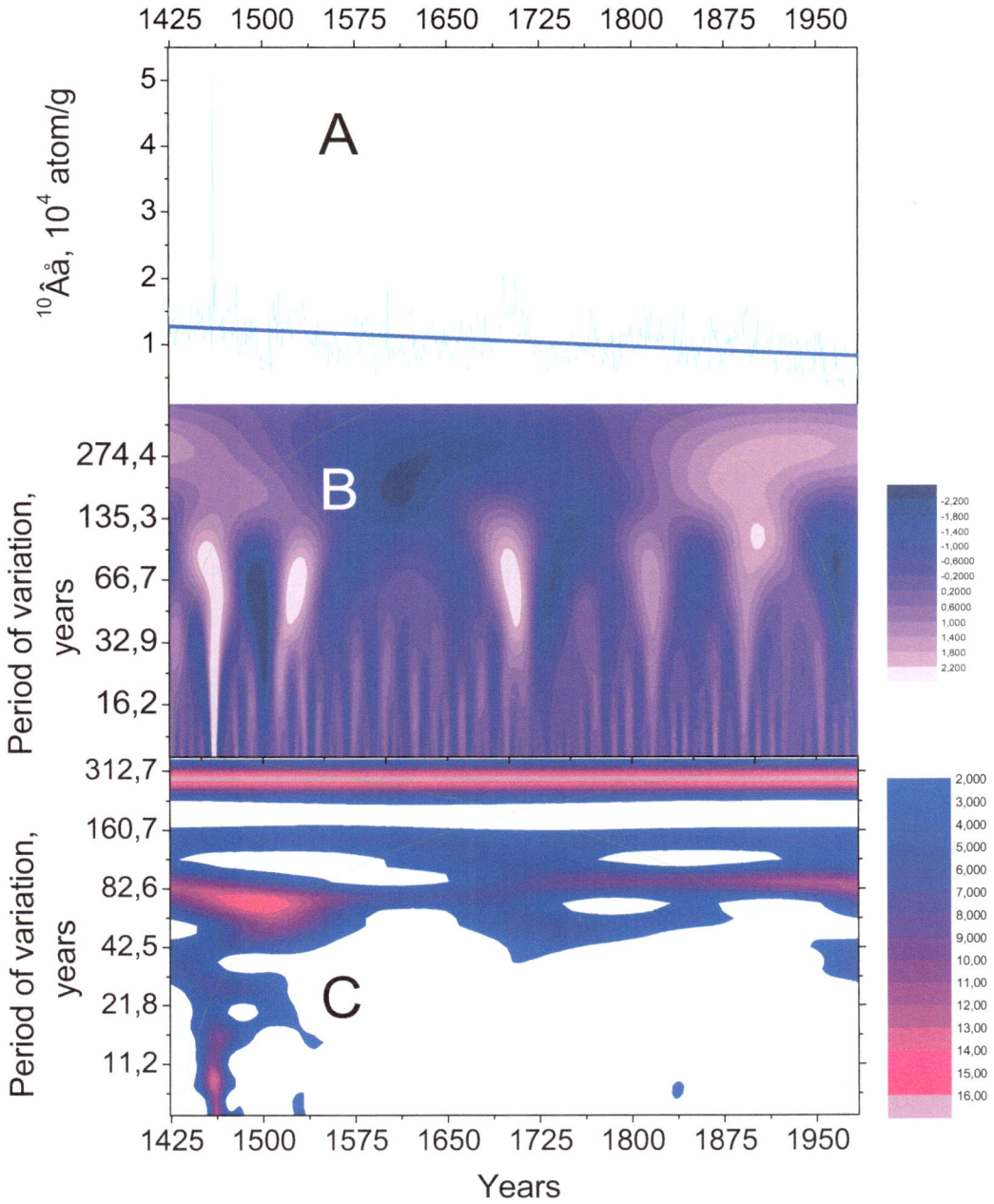

Fig. (2.10). (A) Concentration of ^{10}Be measured in Dye-3 core (South Greenland), blue line - linear trend; **(B)** its local MHAT spectrum, normalized by variance; **(C)** its local Morlet spectrum, normalized by variance. Both spectra were calculated after removing the linear trend. Dashed green lines indicate the cone of influence.

provides a good frequency resolution, while MHAT gives a good resolution in time (according to inequality $\Delta t \cdot \Delta \omega \geq 1/4\pi$ [112]). These properties determine the method of analysis. The Morlet wavelet spectrum is expedient to identify the periods of variation contained in the analyzed signal and the MHAT wavelet spectrum is suitable for the investigation of time evolution of variation found in the previous step.

Wavelet spectra of a cosmogenic beryllium record from the Greenlandic Dye-3 site (65.18 N, 43.83 W, 2480 m a.s.l.) (Fig. **2.10**) covers the period AD 1424-1984 [116, 117]. Dashed green lines indicate the *domain (cone) of influence* - area over which edge effects become important [111] and thus spectrum could be distorted. Wavelet spectra clearly manifest both short-term (quasi 11-year and quasi 22-year) and secular (century-scale and longer) periodicities (Fig. **3.3**).

Using both the Fourier and wavelet approaches is the most effective way to analyse actual solar and geophysical time series. In addition, it is necessary to estimate the statistical significance of the features of the Fourier and wavelet power spectra in order to obtain quantitative estimations of the signal spectral content. This initially requires determination of the type of noises, which distort the analysed time series. It is reasonable to consider the noise present in the geophysical time series as 'red' one. Red noise is a first-order random autoregressive (AR(1) or Markov) process, which has enhanced low-frequency variability arising from the interaction of white noise forcing with the slow-response components of a system:

$$Y(t) = \alpha \cdot Y(t-1) + \varepsilon(t), \tag{2.27}$$

where $\varepsilon(t)$ is the white noise (a set of random numbers with a variance of 1.0) and α represents the autoregression factor.

$$\alpha = \exp(-\frac{1}{\tau}), \tag{2.28}$$

where τ is the characteristic timescale of the AR(1) process (a measure of its 'memory'). The normalized Fourier power spectrum − square of the Fourier transform multiplied by $N/2$ and divided by variance σ^2 − of red noise has a form:

$$P_R^\alpha(\omega) = \frac{1-\alpha^2}{1+\alpha^2 - 2 \cdot \cos(\frac{\omega}{N})} , \tag{2.29}$$

where P_R^α is the spectral power density of red noise with the factor α, ω is a frequency, N is the number of points in the data set. The red noise hypothesis appears quite plausible at least concerning climatic series. Indeed, Hasselmann [118] has suggested a stochastic model which considers the variability of global climate as an integrated response of the ocean-cryosphere-land system to short-scale atmospheric fluctuations ('weather' excitation). The model of Hasselmann predicts 'red' spectra of the internal climatic fluctuations that is in qualitative agreement with the observational data. More complicated models considering processes of convective and diffusive heat exchange in oceans give similar spectra of climatic variability. For example, in the work [119], the following formula for a spectrum of fluctuations of global climate has been obtained:

$$P_{cl\,im}(\omega) = \frac{\sigma_{WN}^2}{\lambda_c^2} \cdot \left\{ \left(1 - \frac{\tau w}{2h} + \frac{\tau}{2h}\sqrt{\frac{v^2+w^2}{2}}\right)^2 + \left(2\pi\omega\tau + \frac{\tau}{2h}\sqrt{\frac{v^2-w^2}{2}}\right)^2 \right\}^{-1} , \tag{2.30}$$

where $P_{cl\,im}$ is the spectral power density, ω is a frequency, σ_{WN}^2 is the white noise variance, λ_c is climatic sensitivity (see Section 6.2), w is the upwelling velocity (see Section 3.2.3), K_z is the vertical diffusion coefficient (see Section 3.2.3), $v^2 = \sqrt{w^4 + (8\pi\omega k_z)^2}$ is the characteristic velocity, h is the depth of the mixed layer of the ocean, and τ is the mean residence time of the global climatic system. If we consider $\sigma_{WN}^2 = 1.0$, $\lambda_c = 0.9$ W m^{-2}/°C, $w = 4$ m yr^{-1}, $h = 100$ m, $k = 4$ m^2 yr^{-1}, and $\tau = 6.3/\lambda_c$ we obtain the spectrum of natural climatic variations (Fig. **2.15**).

Evidently (Fig. **2.11**) the spectrum of inherent fluctuations of the climatic system, calculated according to [119] is well described by red noise with AR(1) factor 0.87. If we assume the noise in the analysed time series $Y(t)$ to be an AR(1) process, statistical significance of the peaks of the Fourier spectrum of $Y(t)$, normalized by variance, can be easily estimated by comparison with the red noise reference spectrum. Notably, confidence level can be calculated as the multiplication of the usual red noise continuum (3.12) by a corresponding χ_2^2 value (see *e.g.* [111]):

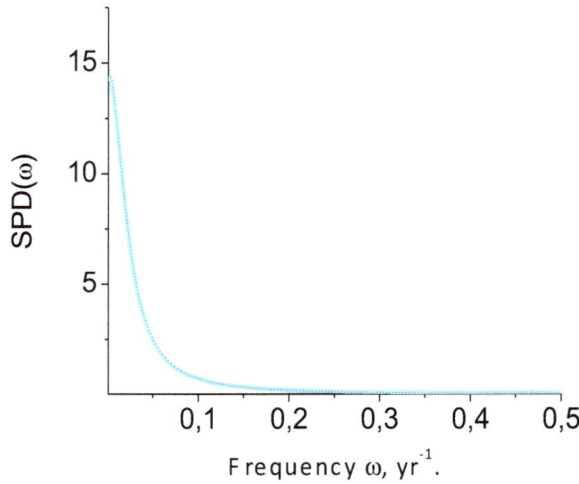

Fig. (2.11). Light cyan line - the spectrum of internal fluctuations of the climatic system, calculated using formula (2.30). Black blue line - the spectrum of red noise with $\alpha=0.87$.

$$P_N^X(\omega) = P_R^\alpha(\omega) \cdot \frac{\chi_{2,X}^2}{2} \qquad (2.31)$$

where $P_N^X(\omega)$ is the spectral power corresponding to confidence level (c.l.) X and $\chi_{2,X}^2$ is the value of chi-squared distribution(χ_2^2) associated with the chosen confidence level X. If a peak in the normalized Fourier spectrum at a given frequency ω_0 reaches the value $P_N^X(\omega_0)$, this peak can be assumed to be a manifestation of a periodic (non-random) signal with a probability X. That is to say, this peak is significant with the confidence level X. Probability of a *null-hypothesis* - the assumption that the peak of the Fourier spectrum is generated by pure random AR(1) process - in that case is $1-X$. This value is also called a *false alarm probability*. In order to analyse the significance of the Fourier spectrum we obviously need to know the value of AR(1) factor α. These factors can be estimated by means of a procedure suggested by [120]. First, the low-frequency component is filtered out of the series under investigation $Y(t)$ and the residual part $Y^{res}(t)$ *is* assumed to consist mainly of noise. Second, the dependence of $Y^{res}(t)$ on $Y^{res}(t-1)$ is plotted and the slope of a linear fit - the coefficient of linear proportionality between $Y^{res}(t)$ and $Y^{res}(t-1)$ - is determined. This value is then considered as an estimation of α. Numerical experiments, made with artificial data sets, show that this procedure allows the determination of the AR(1) factor with an uncertainty usually less than 25% (Fig. **2.12**).

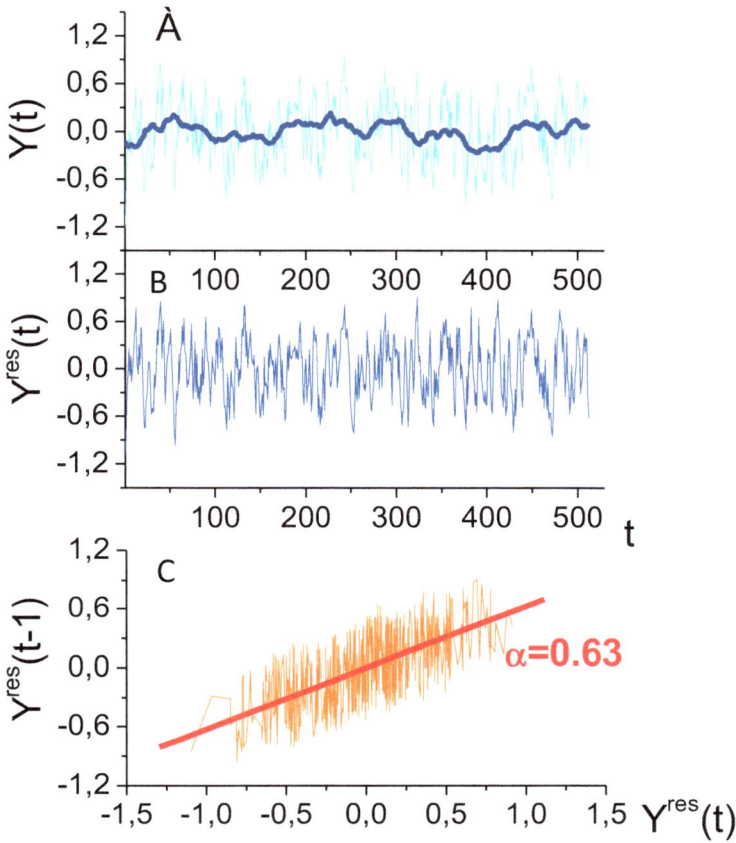

Fig. (2.12). (**A**) Cyan line - $Y(t)$ - red noise with AR(1) coefficient $\alpha=0.6$. Dark-blue line - series smoothed by 40 years moving average; (**B**) series $Y^{res}(t)$ obtained from $Y(t)$ by subtracting the 40 years moving average; (**C**) dependence of $Y^{res}(t)$ on $Y^{res}(t-1)$.

More complicated ways of estimating α, which provide more precise values, are presented in the works [121, 122].

If the noise corrupting the analysed data set is white, *i.e.* $\alpha = 0$ and $\tau = 0$ (no 'memory'), then for an estimation of the significance of a spectral peak at the frequency ω_0 we have to compare it to the possible noise pulses not only at the given frequency ω_0 (the red noise case) but also at all the other frequencies. In that case the significance of the spectrum peaks can be evaluated by means of a formula of Horne and Baliunas [123]:

$$P_F(z) = 1 - \left(1 - \exp(-z)\right)^{\frac{N_i}{2}}, \tag{2.32}$$

where P_F is the false alarm probability, z is the height of the analysed peak of the normalized Fourier spectrum, and N_i is the number of independent frequencies. This number can be derived using the formula of Horne and Baliunas [123]:

$$N_i = -6.362 + 1.193 \cdot N + 0.00098 \cdot N^2, \tag{2.33}$$

where N is a number of points in the analyzed time series. Statistical experiments show that estimations of significance, performed by means of the formula (2.32) give quite correct values at least in the case of a time series consisting of several hundreds or more points. Frescura *et al.* [124] have further improved methods of estimating the significance of the Fourier spectra.

Evaluation of the significance of details of wavelet spectra (estimation of the probability that the detail is not induced by pure noise signal) is an important problem in wavelet analysis. Torrence and Compo [111] have suggested estimating the significance by comparison of wavelet spectrum of an investigated signal with a background reference spectrum. They used theoretically calculated spectra of various noises as background ones. Ogurtsov *et al.* [52] compared wavelet spectrum of the analysed signal with background spectra of test signals, each of which consists of the noise and sinusoid. The amplitude of the sinusoid was chosen so that the corresponding peak in the Fourier spectrum reaches the confidence level X. The wavelet spectra of a number of test signals, containing sinusoids with different frequencies, determine X confidence level for wavelet power. Ogurtsov *et al.* [52] show a method of calculating theoretical 0.99 confidence levels for a time series normalized by variance, which contain 500 and 100 observations and are corrupted by red noises with different AR(1) factors (Fig. **2.13**). Each detail of the local wavelet spectrum, which exceeds the calculated values, could be considered as significant at the 0.99 confidence level.

Finally, a scheme of statistical analysis of solar and geophysical time series could be considered as a consequence of the following steps:

1)   Detrending and pre-whitening - removal of variations with the periods comparable with the length of the analysed time series $Y(t)$. Powerful long-term oscillations can substantially distort Fourier and wavelet spectra, therefore they should be deleted. Subtraction of the polynomial trends of the second and third order is usually enough for this purpose. Detrended signal $Y_1(t) = Y(t) - Y_{tr}(t)$ becomes an object for further analysis.

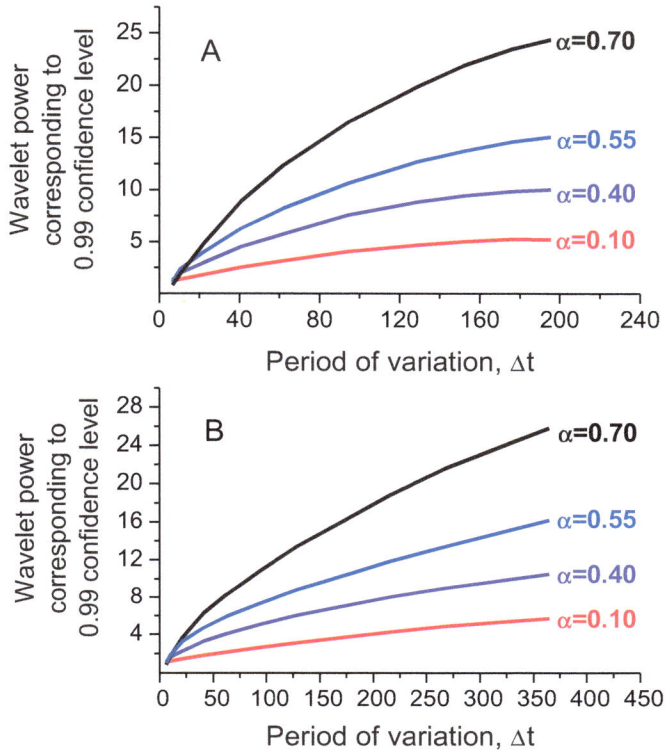

Fig. (2.13). Theoretical 0.99 confidence levels calculated for wavelet spectra (Morlet basis) of time series contain: (**A**) 500 observations; (**B**) 1000 observations. In each case a series was normalized by variance. Δt is the time spacing and α is the AR(1) factor of the noise, which distort the analysed signal.

2)   Calculation of the normalized Fourier spectrum $P_N(\omega)$. The significance of the normalized Fourier spectrum is estimated by the method described above. The use of simple Fourier spectrum in analysis seems to be preferable to calculation of any periodogram by means of a correlation window. Using any correlation window changes the statistics of spectral peaks considerably, which complicates the significance evaluation and often necessitates performing statistical experiments. Another advantage of the Fourier spectrum is the easiness of estimation of amplitudes of the revealed periodic signals. The amplitude can be calculated by integration of the corresponding peak of the normalized spectrum (see Fig. **2.14**) by the formula:

$$A_{mp} = \sqrt{2\ \sigma^2 \int_{\omega_1}^{\omega_2} P(\omega)\ d\omega,}$$ (2.34)

where σ^2 is variance of the analysed signal.

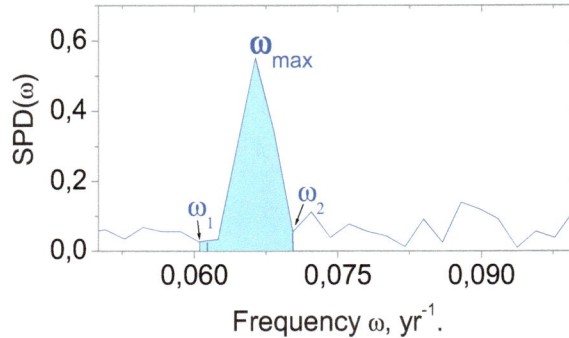

Fig. (2.14). Estimation of the amplitude of a variation with frequency ω_{max} by means of integration of the Fourier spectrum.

3)   Calculation of a local Morlet wavelet spectrum of $Y_1(t)$. The spectrum of Morlet has been used in studies on the evolution of the spectral structure of a signal in time as well as studies on the peculiarities of local non-homogeneities. The significance and details of the normalized Morlet spectrum is estimated by the methods described above. Scalogram, easily determined by means of local wavelet spectrum, can serve as a source of information on a spectrum of the series averaged over the total interval of observations. In comparison with a Fourier spectrum, scalogram is usually more convenient for analysis of long-term variations.

4)   In addition, the method of singular spectrum analysis (SSA) can be applied usefully. The SSA method is based on the decomposition of analysed time series to the eigenvectors of the lag-covariance matrix. These eigenvectors are also called empirical-orthogonal functions (EOFs). The SSA method is aimed at overcoming the problems of finite sample length and noisiness of the studied time series by using a set of data-adaptive basis functions. At first, the scalar time series $Y(t)$ {$t=1, 2,.., N$} is embedded into a vector space of dimension M by constructing M lagged copies $Y(t-j)$ {$j=1, 2,.., M$} thereof. The result is a matrix Z, in which columns contain pieces of the initial series

$Y(t)$ shifted by different lags. Then the $M \times M$ lag-covariance matrix C is calculated and diagonalized:

$$C = \frac{1}{N} Z \cdot Z^T = E \cdot \Lambda \cdot E^T, \qquad (2.35)$$

where Λ is a diagonal matrix of eigenvalues and E is an orthogonal matrix of eigenvectors. The eigenvalue Λ_k $\{k=1,.., M\}$ presents the variance of the time series in the direction specified by the respective eigenvector E_k. Projecting the time series onto each eigenvector (EOF) yields the corresponding temporal principal components (PCs). Linear combinations of the PCs and EOFs (back projection of the PCs onto the eigenvectors) provide the reconstructed components (RCs) - the time series of length N. Summing RCs components of the time series reconstructs it entirely without losing any information. The results of an application of SSA, the annual set of data on the concentration of nitrates (NO_3^-) in the Greenland ice [125], is used as an example (Fig. **2.15**). The signal was decomposed by 15 EOFs, resulting in the four reconstructed components (Fig. **2.15B-D**). The first reconstructed component (RC(1)) describes long-term variations with periods more than 40 years. RC(1) consist of 54% of the total variance. RC(2) and RC(4) describe the quasi five-year periodicity (9% and 7% of variance, respectively). The third component RC(3) expresses decadal (quasi ten-year) periodicity (8% of variance). It is seen from the figure that the general oscillation modes revealed by SSA are distinctly manifested in the Fourier spectrum as well (Fig. **3.8E**). About 15% of the variance falls to the share of spectral peaks in the quasi five-year frequency band. This is in agreement with the 16% obtained by the SSA method. It should be noted that the above amplitude of variation could be easily determined from its variance. It is evident that the SSA approach not only allows the selection of the basic quasi-periodic components which are present in the studied data sets, but also their time evolution over the various timescales to be traced, and the assessment of the amplitudes of the revealed periodicities.

The above scheme can be applied to a solar-terrestrial time series (Fig. **2.16**). An analysis of natural signals, performed using this scheme, makes it possible in difficult cases (strongly non-stationary signals of unusual form) to estimate key parameters of each signal - an average spectrum, its temporal evolution, amplitudes of the variations - by at least two ways. It allows the quality and reliability of the extracted information to be improved considerably and serious errors to be avoided. In the less complicated cases the calculation of the Fourier and Morlet wavelet spectra is sufficient.

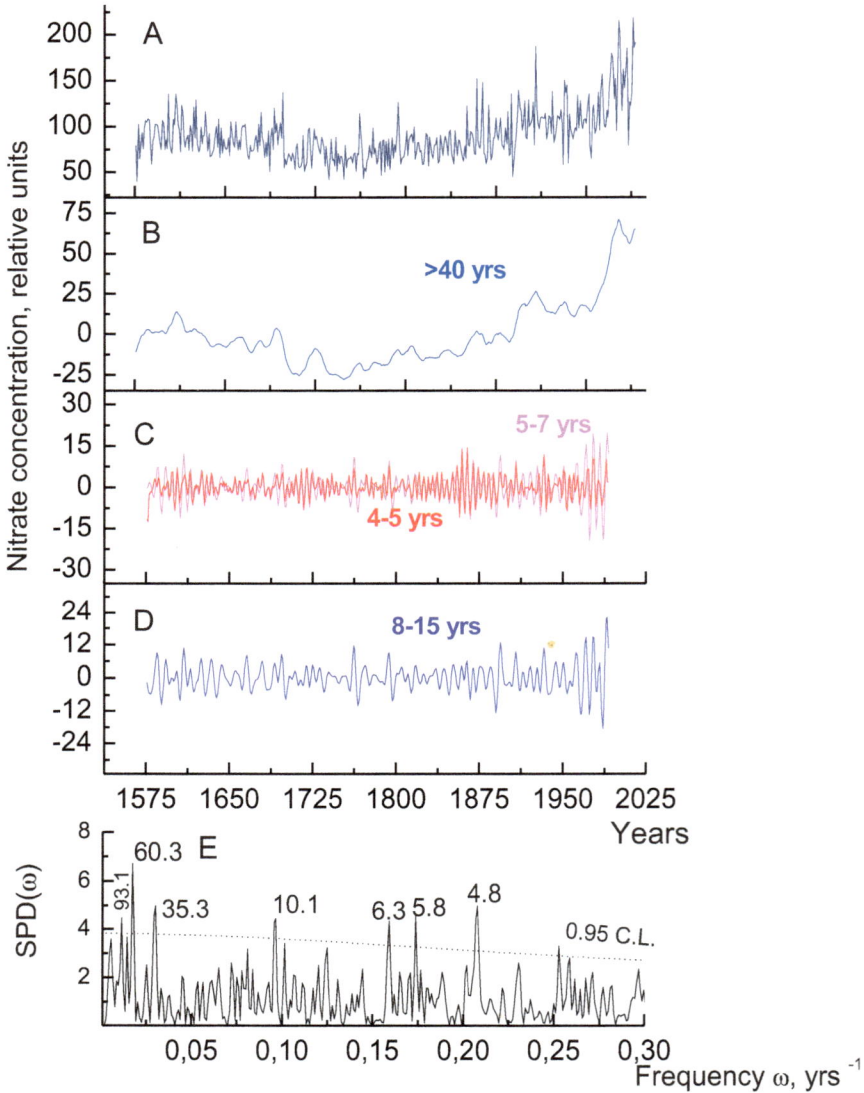

Fig. (2.15). (A) Yearly averaged concentration of nitrate in the Central Greenland ice [125]; **(B)** First reconstructed component of the nitrate series (54% of variance); **(C)** Second (magenta line, 9% of variance) and fourth (red line, 7% of variance) reconstructed components; **(D)** Third reconstructed component (8% of variance); **(E)** Normalized Fourier spectrum of the nitrate data set calculated after subtracting the second-order polynomial trend. Confidence level was calculated for red noise with α=0.12.

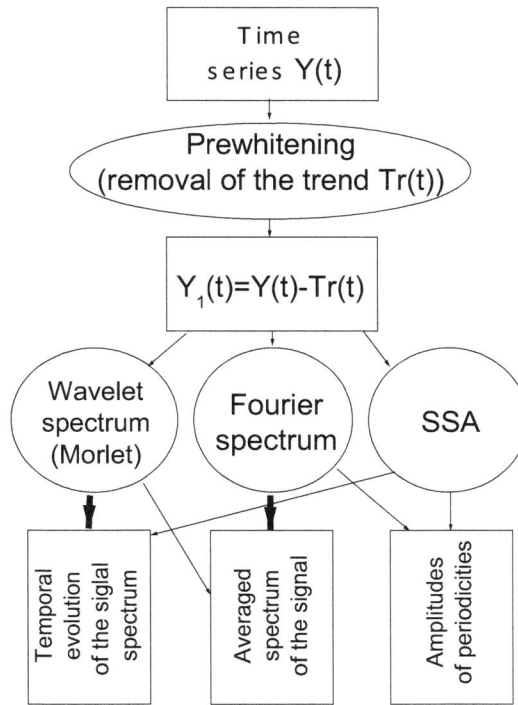

Fig. (2.16). Scheme of statistical analysis of solar, climatic and geophysical time series.

Some other approaches may well also prove useful in investigations of complex data sets. The Hilbert-Huang transform is a new method for analysing non-stationary and non-linear time series [126]. The main point of the method is the "empirical mode decomposition", which decomposes the analysed time series into a limited number of "intrinsic mode functions". With the following Hilbert transform of the "intrinsic mode functions" the signal is presented in an energy-frequency-time distribution (the Hilbert spectrum). The applications of the Hilbert-Huang transform to climate research are described in more detail in [127].

2.7.3. Filtering of Non-Stationary Time Series

It is often necessary to study time variations of signals over certain selected frequency ranges. Methods of filtering should be used for this purpose. Wavelet filtering is a pertinent method and used here for illustration. A time series wavelet filtered in the scale range from a_j to a_k could be calculated by means of the formula of [111]:

$$Y(t,j,k)_{fltr} = \frac{\delta j \cdot \delta t^{1/2}}{C_\delta \cdot \psi(0)} \cdot \sum_{j}^{k} \frac{Re(w(a_j,t))}{\sqrt{a_j}}$$

(2.36)

where C_δ is the reconstruction factor. If the MHAT basis was used, then C_δ =3.541, $\psi(0)$=1.0. The wavelet filtering performed by means of the MHAT basis allows the complicated details of a wide frequency spectrum to be allocated. A simple moving average smoothing as well as state-of-the-art multifractal filtration approaches may also be used for filtering [128, 129]. Multifractal techniques can also be used for filling the gaps [126].

The significance of correlation between wavelet-filtered time series (P_c) is evaluated using a statistical experiment, which includes a number of N Monte Carlo simulations. Every simulation contains:

(a) Generation of random copies of the analysed signals. They were compiled from a Fourier transform of the actual data ($Y(t)$, $Z(t)$), randomizing the phases, and then Fourier transforming back. Artificial series $Y^{rand}(t)$ and $Z^{rand}(t)$, obtained this way, have the same Fourier spectra as the initial ones (see Fig. **2.17**).

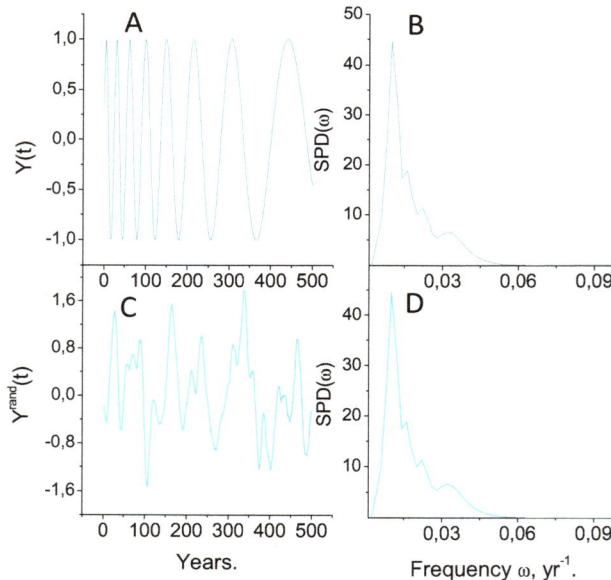

Fig. (2.17). (**A**) time series $Y(t)$; (**B**) its Fourier spectrum; (**C**) $Y^{rand}(t)$ - a random copy of $Y(t)$, obtained by randomization of phases of the Fourier transform; (**D**) Fourier spectrum of $Y^{rand}(t)$.

(b)   Wavelet filtering of the random copies. Time series $Y_{fltr}^{rand}(t)$ и $Z_{fltr}^{rand}(t)$ are calculated at this step.

(c)   Calculation of the correlation coefficients between the two filtered random copies R_i.

(d)   Its comparison with the coefficient of correlation between the filtered actual data sets R_0 (see [48] for more detail).

Then K^+ - the number of cases when $|R_i| > |R_0|$ - is estimated and significance is defined as $P_c = 1 - K^+/N$. As noted by Traversi *et al.* [130] the random-phase test tends to underestimate the confidence level and thus it is reasonable to consider the obtained significance estimations as rather conservative or as only the lower limits (Fig. **2.18**).

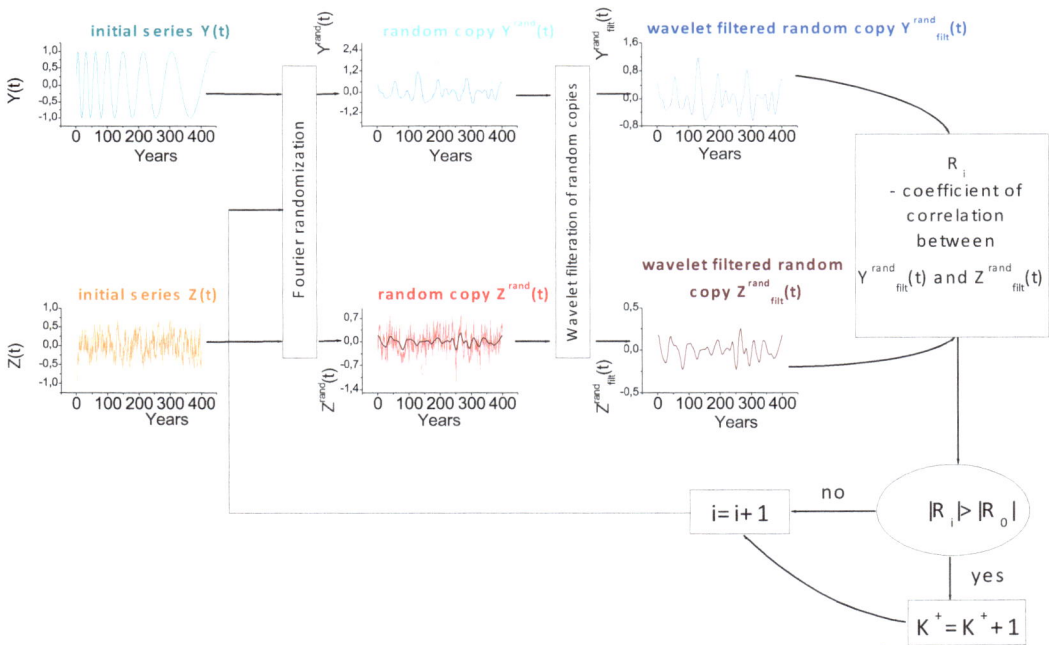

Fig. (2.18). Scheme of the statistical experiment performed for evaluation of the significance of the correlation between two filtered time series $Y(t)$ and $Z(t)$.

As we have shown in this Chapter both modern heliophycics and solar-terrestrial physics as well as statistical analysis of time series achieved remarkable progress during the last decades. The shortness of the available instrumental records is an important limiting factor. The study of paleoproxies, spanning time intervals from centuries to millennia, allows us to overcome this difficulty.

CHAPTER 3

Solar Paleoastrophysics: Advances and Limitations

Abstract: In addition to a historical overview and general information, major advances and problems of solar paleoastrophysics are described, including the data obtained by means of cosmogenic isotopes (^{14}C, ^{10}Be), nitrates (NO_3^- ion), and historical observations. The most recent proxy-based paleoreconstructions of sunspot numbers and sunspot area, covering time intervals from 150 to 10 000 years, are presented. Possible sources of time variations of the concentration of ^{14}C, ^{10}Be and NO_3^- ion in terrestrial archives are considered.

Keywords: Cosmogenic isotopes, nitrate, solar paleoastrophysics,

Paleoastrophysics is a science studying astrophysical phenomena which occurred prior to the times of instrumental astronomy. Solar paleoastrophysics uses both the of historical data (catalogues of aurorae and sunspots as well as unaided eye observations) and proxy indicators of solar activity (the abundance of cosmogenic isotope and nitrate in terrestrial archives). The basic ideas of paleoastrophysics were formulated by B.P. Konstantinov and G.E. Kocharov [131] in the framework of a program called 'Astrophysical phenomena and radiocarbon'. They demonstrated that it is possible to obtain reliable information on a number of astrophysical phenomena of the distant past using various natural archives. The phenomena include long-term variations in the intensity of solar and galactic cosmic rays, magnetic and flaring activity of the Sun, supernova outbursts, and gamma-ray bursts. The first series of high-precision radiocarbon measurements with good time resolution (1-2 years) were made in 1980s [132-135]. Currently, paleoastrophysics has become an important branch of science producing increasingly significant results.

3.1. Cosmogenic Isotopes - Mechanisms of Formation and Fixation in Terrestrial Archives

Studies on the relative abundance of cosmogenic isotopes in natural archives are among the main tools of solar paleoastrophysics. Radiocarbon ^{14}C and radioberyllium ^{10}Be are produced in the Earth's atmosphere due to the effect of high-energy GCR. About two thirds of ^{10}Be and ^{14}C originate in the stratosphere and the remaining part is produced in the troposphere.

Maxim Ogurtsov, Risto Jalkanen, Markus Lindholm and Svetlana Veretenenko

3.1.1. Cosmogenic ^{14}C

Cosmic particles with energy more than 1 GeV/nucleon are the major sources of ^{14}C generation. The main part of radiocarbon is generated by secondary thermal neutrons in a reaction $^{14}N(n,p)^{14}C$ having a large cross-section at thermal neutron energies and a resonance at energies of an order of several MeV [135]. Thermal neutrons present in the atmosphere are a product of the cosmic-ray-induced cascade. The contribution of other reactions - $^{16}O(n,^{3}He)^{14}C$, $^{15}N(n,d)^{14}C$, $^{17}O(n,\alpha)^{14}C$ - is rather weak. The rate of neutron production in the atmosphere depends on variation in cosmic ray intensity. Assessments of the modern global average rate of production of ^{14}C in the atmosphere range from 1.6-1.8 at$\times$cm$^{-2}\times$s^{-1} [136, 137] to 1.9-2.5 at$\times$cm$^{-2}\times$s^{-1} [138, 139]. This means that ca 3×10^{26} atoms of ^{14}C are generated in the atmosphere during a year. The total mass of ^{14}C in the whole atmosphere is 45-75 tons. Kocharov *et al.* [140] showed that during 1956-1972 the contribution of SCR to the production of ^{14}C was only 10-15% of the GCR contribution. Later estimations of Kovaltsov *et al.* [136] suggested much lower values: 0.8% of the total ^{14}C production over 1954-1965 and 70% of the monthly mean contribution due to a solar proton event on 23 February 1956. After origination ^{14}C generated in the atmosphere is oxidized rapidly to ^{14}CO:

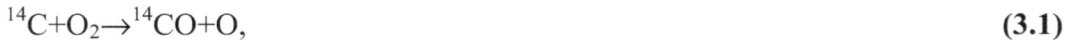

$$^{14}C+O_2\rightarrow{}^{14}CO+O, \tag{3.1}$$

and after 1.0-1.5 months to $^{14}CO_2$ [141]:

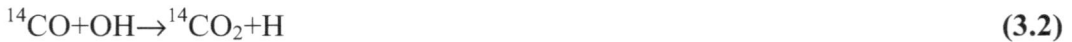

$$^{14}CO+OH\rightarrow{}^{14}CO_2+H \tag{3.2}$$

which, in turn, is mixed in the global atmospheric $^{12}CO_2$ pool, involved in the global carbon cycle - a chain of geochemical and geophysical processes - and finally fixed in tree rings- see Fig. (3.1).

Radiocarbon decays with a half-life of 5730 years, that makes it possible to study phenomena that happened up to 80 000 years ago [142]. Since the flux of GCR is governed by solar activity, the rate of ^{14}C generation is also modulated by the activity of the Sun. This modulation can be described by formula:

$$Q=Q_0-\frac{dQ}{dW}\times W, \tag{3.3}$$

where Q is in at$\times$cm$^{-2}\times$s^{-1} and W is Wolf number; for estimations of Q_0, and $\dfrac{dQ}{dW}$,

see Table **3.1**.

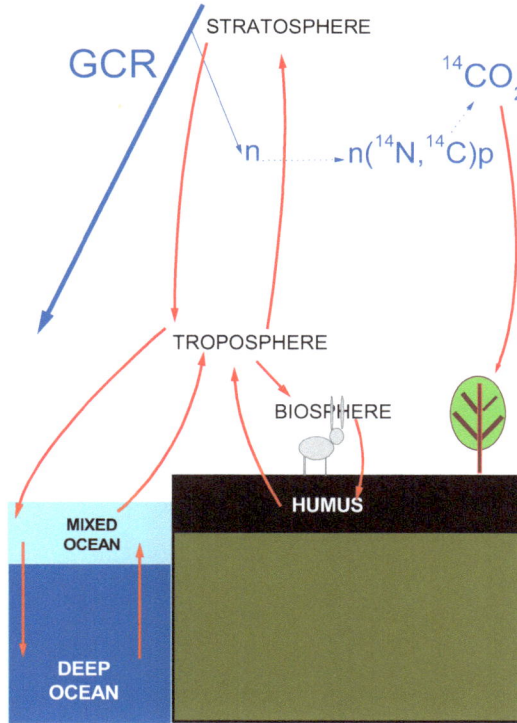

Fig. (3.1). Circulation of radiocarbon in carbon-exchange system.

Table 3.1. Different estimations of the parameters Q_0 and $\dfrac{dQ}{dW}$.

	Time Interval of Estimation	Q_0 (at$\times$cm$^{-2}\times$s^{-1})	$\dfrac{dQ}{dW}$ (at$\times$cm$^{-2}\times$s^{-1})
O'Brien [137]	1937-1970	1.937	2.42×10^{-3}
Korff and Mendell [143]	1965-1975	2.60	5.53×10^{-3}
Stuiver and Quay [144]	20th century	2.091	4.10×10^{-3}
Kocharov *et al.* [145]	20th century	2.559	7.00×10^{-3}

According to different estimations, the peak to trough change of the rate of radiocarbon generation over a solar cycle reaches 20-30%. The dependence of Q on GCR intensity can be calculated using the formulae of Allan and Lowe [141]:

$$Q = 4.03\text{-}4.67 \times 10^{-3}\, \Phi + 4.96 \times 10^{-6}\, \Phi^2 - 2.33 \times 10^{-9}\, \Phi^3,$$

$$\Phi = -22831 + 21.503\, N - 6.1312 \times 10^{-3}\, N^2 + 5.5673 \times 10^{-7}\, N^3, \tag{3.4}$$

where Q is in at×см$^{-2}$× s^{-1}, N, representing the hourly Climax neutron counting rate (see Fig. **2C** of Chapter 2) in units 100 counts/hour and Φ is solar modulation parameter - the value shows how much energy a cosmic ray particle must have to pass through the heliospheric magnetic field. It is expressed in MV. In solar cycle minima, Φ is ca 400 MV while in maxima it can reach more than 1000 MV. The average Φ since 1950 is ca 800 MV. Castagnoli and Lal [146] suggested an easier formula for the calculation of Q:

$$Q = 3.2\text{-}1.2 \times 10^{-3}\, \Phi \tag{3.5}$$

Kocharov *et al.* [147] proposed the formula:

$$Q = \left(\frac{I}{385}\right)^{0.4082} \tag{3.6}$$

where Q is in at×cm^{-2}×s^{-1} and I is the cosmic-ray intensity in the energy range 0.2 to 50 GeV in m^{-2}×s^{-1}× ster^{-1}.

After generation, $^{14}CO_2$ takes part in the global carbon cycle (see Fig. **3.1**) and during photosynthesis it is finally fixed in tree rings. The wood of tree rings is composed of cellulose, hemicelluloses, lignin, extractives, fats, and resin. Cellulose contributes to 40-55% of the wood's dry weight, while hemicellulose (20-25%) and lignin (20-30%) compose the remainder of that weight. Cellulose and hemicellulose are polysaccharides while lignin is a group of a complex natural polymers. Extractives - sugars, phenols and starch - are soluble in hot water and give wood its taste and smell. They constitute about 3-5% of dry wood. Fat, resin, and wax, which allow the migration of carbon between tree rings, contribute 3-4% of wood weight. Extractives and lignin could also play some role in carbon migration. Carbon exchange between tree-ring cellulose is practically absent. However, the migration of carbon between tree rings can affect concentration appreciably - see Fig. (**3.1**), which draws on the work of Stuiver and Quay [148].

For this reason, all fractions of wood, except cellulose, are removed at the stage of the preliminary chemical preparation of samples. If a chemical treatment is carried out accurately, the retained cellulose becomes a reliable natural archive of radiocarbon.

The life time of radiocarbon in the atmosphere is 4-10 years [149]. It should be noted that in relation to ^{14}C, the term "life time" is conditional. In this case, it has some characteristics of the rate of change of isotope content in the atmosphere - the physical process, which is not equivalent to processes of natural carbon exchange. The life time of the actual CO_2 molecule in the atmosphere is 50-200 years according to IPCC [150]. Because of this rather lengthy time in residence, $^{14}CO_2$ is well mixed in the Earth's atmosphere prior to deposition in tree rings. Thus latitudinal effects in ^{14}C concentration are effectively smoothed out. Radiocarbon in tree rings is sensitive to changes in the geomagnetic dipole field, which shields terrestrial atmosphere from cosmic particles at low and middle latitudes. The dependence of Q on geomagnetic field is described by formula:

$$Q(t) = Q_0 \left(M(t) / M(0) \right)^{-\gamma} \tag{3.7}$$

where $M(t)$ is intensity of geomagnetic field at time t, $M(0)$ is the contemporary value of geomagnetic field intensity ($\cong 8 \times 10^{22}$ A$\times$m^2), Q_0 is the modern value of production rate, and $\gamma = 0.50$-0.63 [135, 151]. A weaker geomagnetic field causes an increase of cosmic ray flux reaching the Earth's atmosphere and results in higher rate of ^{14}C generation. Consequently, the oscillation of radiocarbon concentration in tree rings should reflect either the variation of the rate of generation in the atmosphere or the variation in the carbon-exchange cycle. Changes of the rate of ^{14}C generation in the atmosphere appear as a result of the corresponding variations of solar activity and geomagnetic dipole intensity. Changes in the carbon recirculation cycle could be caused by variations of global climate. Local climatic variability is usually presumed to play a minor role in variations of atmospheric radiocarbon due to the global mixing of CO_2. Methods of dendrochronology are useful for exact dating with annual resolution of the samples.

^{14}C decays by emitting electrons (β-decay) with maximum energy of 156 KeV. Radiocarbon concentrations in wood samples are measured by means of the registration of β-particles using proportional gas counters, liquid scintillation detectors, and accelerator mass spectrometers. The analyzed sample is converted to gases (CO_2, CH_4, C_2H_2) if a proportional counter is used and to the liquid benzene and etilbenzene if the scintillation detector is used. One gram of wood sample usually contains ca 6×10^{10} atoms of ^{14}C. Tree-ring $\delta^{14}C$ levels are determined as relative deviations in ^{14}C activity from the NBS oxalic acid standard standard $\left(^{14}C/^{12}C\right)_{NBS} = 1.176 \times 10^{-12}$:

$$\delta^{14}C = (1 - \frac{A_S}{A_{NBS}}) \cdot 1000, \qquad (3.8)$$

where $\delta^{14}C$ is measured in per mille, A_S is the activity of the sample, and A_{NBS} is the activity of the standard (14.27±0.7 decays/(minute×g of carbon)). The value $\delta^{14}C$ is then corrected for an isotopic fractionation effect:

$$\Delta^{14}C = \delta^{14}C - 2.0 \cdot (\delta^{13}C + 25) \cdot (1 + \frac{\delta^{14}C}{1000}) \qquad (3.9)$$

where $\delta^{14}C$, $\delta^{13}C$ are concentrations of ^{14}C and ^{13}C measured in the experiment ($\delta^{13}C$ is the per mille deviation from ^{13}C content in the standard belemnite sample), $\Delta^{14}C$ is measured in per mille. Coefficient 2.0 in the correction formula was established by Stuiver and Polach [152] and is now standard, despite other values obtained by different researchers: 1.85-2.05 [153, 154]. Deviations of the correction coefficient from 2.0 could add some uncertainty to the radiocarbon age of the sample [153], but the discrepancy is unlikely very important for paleoastrophysics. Modern experimental installations measure radiocarbon concentration with precision up to 2 per mille (0.2%). The most up to date IntCal09 and Marine09 radiocarbon record extends back in time to 50,000 BP (years before AD 1950) [155].

3.1.2. Cosmogenic ^{10}Be

Cosmogenic ^{10}Be is produced in the Earth's atmosphere through nuclear spallation reactions $^{14}N(Ha,X)^{10}Be$ and $^{16}O(Ha,X)^{10}Be$, where Ha are hadrons of the atmospheric cascade, induced by cosmic-ray particles, and X are the other products of the reaction. The reactions have a threshold character; *i.e.* they can occur only if the hadron energy exceeds ca 15 MeV. The mean production rate of cosmogenic beryllium in the atmosphere is $(2.0-2.7) \times 10^{-2}$ at×cm^{-2}×s^{-1}. ^{10}Be oxidizes quickly to ^{10}BeO and it is then is absorbed by aerosols (see Fig. **3.2**).

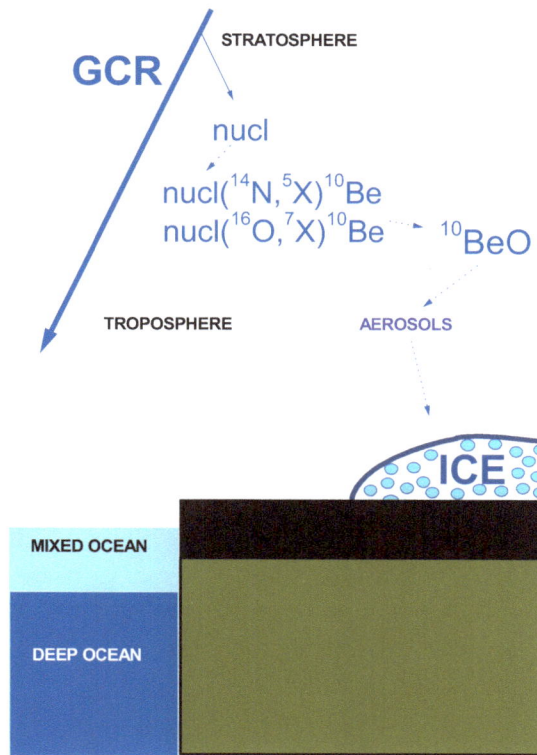

Fig. (3.2). Generation of a ^{10}Be record in polar ice.

The stratospheric transport of ^{10}BeO is determined by gravitational settling and advection. In the troposphere cosmogenic beryllium is involved in processes of wet (associated with precipitation) and dry deposition and finally is preserved in polar ice and sea-bottom deposits. Dry deposition is connected with sedimentation and is not affected by variations in precipitation. Wet deposition is generally dominant. It contributes for more than 90% of a global deposition value [156] while dry deposition could be important over domains with very low precipitation (see [157]). For example, in central Antarctica it can reach 60-80% of the total value. The mean residence time of ^{10}Be in the stratosphere is 1-2 years and in the troposphere ca 1 month [142, 158]. Accordingly, ^{10}Be produced in the troposphere has no appreciable latitudinal transfer and it is deposited mostly locally. Partly or totally mixed stratospheric ^{10}Be is deposited mainly at subtropical latitudes - over the area of tropopause breaks where an exchange between the troposphere and the stratosphere is facilitated (see Fig. **4** after Heikkilä *et al.* [156]). It was suggested earlier that beryllium, which is deposited at high (>70°) latitudes, has mainly a tropospheric origin [149]. Steig *et al.* [159] also assumed that most of the ^{10}Be deposited in Central Antarctica is of local origin. However, more recent calculations of Heikkilä *et al.* [160] showed that 60% of ^{10}Be in Greenland and 70% in Antarctica have stratospheric (thus, well-mixed) origin. Since the reactions leading to ^{10}Be generation have a threshold character, at some latitudes the energy of geomagnetic cut-off becomes less than the threshold energy. In this case, the influence of geomagnetic fields on the process of beryllium generation disappears. In modern values, this dipole moment entails a latitude close to 68°. The dependence of global production rate of beryllium Q on geomagnetic field calculated by Lal [161] can be roughly approximated by formula:

$$Q(t) = Q_0 \left(0.57 + 1.46 \cdot \exp(-1.224 \frac{M(t)}{M(0)}) \right), \tag{3.10}$$

where $M(t)$ is the intensity of geomagnetic field at time t, $M(0)$ is the modern value of geomagnetic field intensity, and Q_0 is the contemporary value of production rate. Because of this limited time, ^{10}Be stays in the atmosphere and the simplicity of the way beryllium is delivered to natural archives its eventual concentration in ice, is almost linearly connected with the rate of generation in atmosphere. It is described by the formula:

$$\left[^{10}Be \right] = \frac{^{10}F(at \times cm^{-2} \times year^{-1})}{A(g \times cm^{-2} \times year^{-1})}, \tag{3.11}$$

where ^{10}F is the flux of beryllium and A is the rate of ice accumulation. The rate of ice generation is determined by precipitation and it is therefore dependent on local meteorological conditions. The flux of ^{10}Be is influenced by variations in tropospheric inter-latitude mixing and exchange between the stratosphere and troposphere. That is why fluctuations of local and regional meteorological parameters could affect both short-term (periods less than 10 year) [116]) and longer-term (multidecadal and longer periods) [162, 163] variations of beryllium concentration in ice. The connection between the measured ^{10}Be concentrations and fluxes in ice cores and the global production rate is not yet completely understood [160].

Beryllium concentration is measured in stratified and dated ice cores. The dating of ice cores is achieved by means of stratigraphic and isotope methods. Usually it is possible to discriminate annual layers at least at the upper part of the core. During the warm period temperatures are higher and ice crystals and grains can grow to larger sizes. Therefore summer ice contains a lot of air. During cold season crystals and grains are smaller and densely packed by winds which are stronger in winter. Thus winter ice contains less air and is denser. Summer layers of ice and firn look whiter while winter layers are darker and more transparent. Moreover, if ice or firn melts during the warm period and the drain of melted snow is absent, subsequent freezing forms ice crust, which can be easily distinguished. The dust content of summer and winter layers also quite often differs. As a result, a depth-age relationship for polar firn can be determined by means of optical stratigraphy [163]. When isotope methods are applied in dating, for example, the concentration of a stable isotope ^{18}O can be measured in ice samples in parallel with ^{10}Be. The concentration of ^{18}O has a distinct annual cycle which helps precise dating. The concentration of H_2O_2 was measured simultaneously with ^{10}Be in ice core retrieved from the Dye-3 station in Southern Greenland [116]. Thus analysis of seasonal cycles in isotope concentration and visible annual layers, performed together with the detection of volcanic horizons of known age (using sulfate concentration anomalies measured in the ice). A joint application of stratigraphic and isotope methods usually also yields good results. However, over the larger depths, the differences between layers are smoothed out due to high pressure. Thus it is necessary to use modeling representations of ice spread for dating more ancient ice (aged over several thousand years). The rate of snow accumulation, the distance of a core from a watershed, the thickness of an ice sheet, and the data of mass balance accounting for vertical velocity serve as initial data at modeling. Nevertheless, modern approaches allow researchers to provide dating of cores from Central Greenland with the maximum error of the time scale close to 5% in the last Glacial and 1% in the Holocene. Thus, an absolute uncertainty of dating is about 1.3 kyr at 60 kyr ago [164]. Among the beryllium records currently available, some have the time resolution of 1 year [116, 117, 165].

For ^{10}Be decays with a half-life of 1.5×10^6 years, this is long enough to study astrophysical processes with time scales of few millions of years. Because of the long life time, the beryllium concentration is very low and is difficult to measure by the decay rate. McCorkell *et al.* [166] had to use 1200 tons of water obtained from melted ice for such a measurement. However, Raisbeck *et al.* [167] constructed a ^{10}Be time series from only 10 kg of water by means of accelerator mass-spectrometry (AMS). Currently, the AMS method is applicable if an amount of at least1 10^7 atoms of cosmogenic beryllium is available. Its concentration often is defined in units of 10^4 atoms/g. The precision of determining annual values of ^{10}Be concentration is 4-10% [116].

Thus, the abundance of cosmogenic ^{14}C and ^{10}Be in terrestrial archives should reflect time variations of:

(a)   solar magnetic activity;

(b)   the flux or the shape of GCR spectrum outside the of solar system;

(c)   intensity of a dipole geomagnetic field of the Earth;

(d)   a global and regional climate;

Moreover, ^{10}Be records could be influenced by local meteorology.

It is obvious that, despite the potential contribution of many factors to the concentration of cosmogenic isotopes (Table **3.2**), it is a valuable source of information on past fluxes of galactic cosmic rays and solar activity.

Table 3.2. Main characteristics of cosmogenic isotopes.

	Rate of Generation (at $cm^{-2} s^{-1}$)	Main Reactions of Generation	Life Time Towards Decay (Years)	Energy of Particles Generating the Isotope (GeV)	Residence Time in Atmosphere (Years)	Natural Archive	Latitu-Dinal Effect
^{14}C	2.6-2.5	^{14}N(n,p)^{14}C	5.7×10^3	0.2-50	3-4	Tree rings	Absent
^{10}Be	$\approx 2.0 \times 10^{-2}$	^{14}N$(A,X)^{10}$Be, ^{16}O$(A,X)^{10}$Be	1.5×10^6	$\approx$1-10	0.1-1	Polar ice, bottom sediment	Present

3.2. Variations of ^{14}C in Tree Rings and their Paleoastrophysical Implementations

The reconstruction of the radiocarbon production rate Q from the measured Δ^{14}C value is not straightforward because of the properties of the carbon-exchange system.

3.2.1. Carbon Exchange Cycle and Reconstruction of Solar Activity Using ^{14}C Data

The process of radiocarbon re-circulation can be described by many-reservoir carbon exchange models. Box models have been applied in many works (*e.g.*, [168-170]) they provide quite an adequate description of ^{14}C variation over the last few centuries. In a box-model, the circulation of ^{14}C is represented by fluxes between some well mixed and quasi-homogeneous carbon reservoirs. Kocharov *et al.* [168] used a simple 5-reservoir model (Fig. **3.3**).

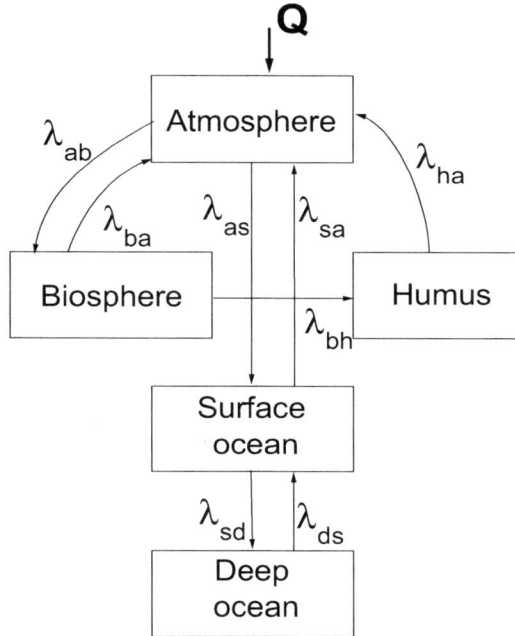

Fig. (3.3). A model of the carbon exchange system based on five reservoir.

The corresponding system of differential equations is:

$$\dot{N}_a = Q(t) - (\lambda + \lambda_{ab} + \lambda_{as})N_a + \lambda_{ba}N_b + \lambda_{ha}N_h + \lambda_{sa}N_s,$$

$$\dot{N}_b = \lambda_{ab}N_a - (\lambda + \lambda_{ba} + \lambda_{bh})N_b,$$

$$\dot{N}_h = \lambda_{bh}N_b - (\lambda + \lambda_{ha})N_h,$$

$$\dot{N}_s = \lambda_{as}N_a - (\lambda + \lambda_{sa} + \lambda_{sd})N_s + \lambda_{ds}N_d,$$

$$\dot{N}_d = \lambda_{sd}N_s - (\lambda + \lambda_{ds})N_d.$$

$$(3.12)$$

where indices *a, b, h, s, and d* denote main carbon reservoirs - the atmosphere, biosphere, humus, surface (mixed) layer of ocean (usually 50-200 top meters), and deep layer of ocean. Likewise, N_i indicates the concentration of radiocarbon in i reservoir; $\lambda_{ij} = 1/\tau_{ij}$ are the rates of exchange between i and j reservoirs; $\lambda = 1/5370$ years is the rate of radiocarbon decay; and $Q(t)$ is a global rate of radiocarbon production in the atmosphere. In order to interpret experimental data; that is, to calculate $Q(t)$ from N_a - measured in the concentration of ^{14}C in tree rings - one should solve the equations (4.12). The solutions lead to the following system:

$$N_b(t) = \exp(-(\lambda + \lambda_{ba} + \lambda_{bh})t) \cdot \left[N_b(0) + \int_0^t \lambda_{ab} N_a(x) \exp(-(\lambda + \lambda_{ba} + \lambda_{bh})x)dx \right],$$

$$N_h(t) = \exp(-(\lambda + \lambda_{ha})t) \cdot \left[N_h(0) + \int_0^t \lambda_{bh} N_b(x) \exp(-(\lambda + \lambda_{ha})x)dx \right],$$

$$N_o(t) = -N_o(0)\left(\frac{r_2}{r_1 - r_2}\right)\exp(r_1 t) + N_o(0)\left(\frac{r_1}{r_1 - r_2}\right)\exp(r_2 t) +$$

$$+ \int_0^t \lambda_{sd} \lambda_{as} N_a(x)\frac{\exp(r_2(t-x)) - \exp(r_1(t-x))}{r_2 - r_1}dx,$$

$$N_s(t) = N_s(0)\exp(-(\lambda + \lambda_{sa} + \lambda_{sd})t) + \int_0^t \exp[(\lambda + \lambda_{sa} + \lambda_{sd})(x-t)] \cdot (\lambda_{as}N_a(x) - \lambda_{ds}N_d(x))dx,$$

$$Q(t) = \overset{\bullet}{N}_a(t) + (\lambda + \lambda_{ab} + \lambda_{as})N_a(t) - \lambda_{ba}N_b(t) - \lambda_{ha}N_h(t) - \lambda_{sa}N_s(t),$$

$$r_1 = 0.5 \cdot \sqrt{(\lambda_{ds} + \lambda_{sd} + \lambda_{sa})^2 - 4 \cdot \lambda_{ds}\lambda_{sa}} - 0.5 \cdot (2\lambda + \lambda_{ds} + \lambda_{sd} + \lambda_{sa}),$$

$$r_2 = -0.5 \cdot \sqrt{(\lambda_{ds} + \lambda_{sd} + \lambda_{sa})^2 - 4 \cdot \lambda_{ds}\lambda_{sa}} - 0.5 \cdot (2\lambda + \lambda_{ds} + \lambda_{sd} + \lambda_{sa}) \qquad \textbf{(3.13)}$$

$N_i(0)$ here are equilibrium concentrations. $N_a(t)$ could be determined from the measured $\Delta^{14}C$ values. If we calculate all integrals numerically, using parameters:

τ_{ab}= 30 yrs, τ_{as}= 9.45 yrs, τ_{ba}= 28 yrs, τ_{bh}= 40 yrs, τ_{ha}= 120 yrs, τ_{sa}= 21.3 yrs, τ_{sd}= 21.3 yrs, τ_{ds}= 900 yrs, $N_a(0)$ = 5.7×10^9 at×cm^{-2}, $N_b(0)$ = 3.27×10^9 at×cm^{-2}, $N_h(0)$ = 9.22×10^9 at×cm^{-2}, $N_s(0)$ = 1.2×10^{10} at×cm^{-2}, and $N_d(0)$ = 4.095×10^{11} at×cm^{-2},   **(3.14)**

we determine $Q(t)$. Then we can calculate the group sunspot number R_Z using the formula, applied by Korff and Mendel [143], for Wolf numbers:

$$R_Z(t) = 181.5 \ (2.6 - Q(t)), \qquad\qquad\qquad\qquad\qquad\qquad \textbf{(3.15)}$$

And, finally, this allows us to obtain the radiocarbon based group sunspot number reconstruction (Fig. **3.4**). It is shown together with group sunspot number, measured instrumentally, smoothed by 11 years, and interpolated by decades.

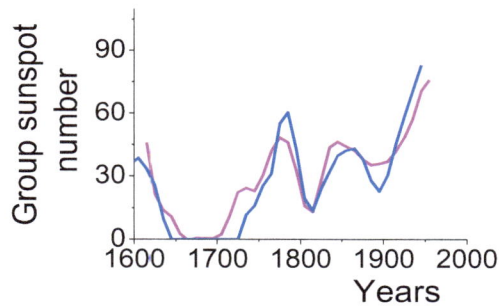

Fig. (3.4). Magenta curve - actual observed group sunspot number [37], blue curve - sunspot number reconstructed from the decadal $\Delta^{14}C$ series [171].

Data on $\Delta^{14}C$ were taken from Stuiver *et al.* [171] and corrected for industrial Suess-effect. The coefficient of correlation between actual (derived from telescopic observations) and reconstructed sunspot data is 0.92 in AD 1615-1945 while the standard deviation is only 8.3. Parameters (3.14) were determined using their rough estimations [172-174] and following fitting the reconstructed sunspot numbers to actual values of group sunspot number. Approximate assessments of the equilibrium concentrations of radiocarbon in reservoirs and rates of inter-reservoir exchange were obtained using the experimental ^{14}C measurements and observation of: (a) the anthropogenic dilution of the atmospheric ^{14}C over the 19th-20th centuries (the effect of Suess); (b) the distribution of artificial ^{14}C, produced by atmospheric nuclear detonation (bomb-effect) between reservoirs. During the nuclear bomb tests of late 1950s and 1960s, 3×10^{27} to 3×10^{28} atoms of radiocarbon were injected to the atmosphere yearly [175]. As a result, the radiocarbon concentration in the Northern Hemisphere nearly doubled by the beginning of the 1960s.

The radiocarbon-based solar reconstructions of many other authors (*e.g.* [133, 176]) also delineate the actual sunspot numbers quite precisely. During AD 1645-1715 - the time period of the Maunder minimum (MM) - calculated sunspot numbers appear as less than zero. It is a result of a strong rise in ^{14}C concentration through the MM, which exceeds the value that could be reached if the sunspot number is equal to zero. In the present reconstruction, negative values were equated to zero. Stuiver and Quay [144] assumed that during the deep minima of the Sun's activity, sunspot numbers were not only close to zero so, too, was geomagnetic activity. It is in contrast with the contemporary period, when space weather has been far from fully calm and even sunspots are absent. Currently, during minima of solar cycles ($R_Z\cong0$), the solar wind continues to blow and geomagnetic *Aa* index still is considerably distinct from zero. Damon and Sonett [177] consider that a very high concentration of ^{14}C in the second half of the 17th century is connected with additional climatic contributions. According to Damon and Sonett [177], a decrease of solar activity during MM has caused a reduction of luminosity of the Sun, a global cold snap, and the cooling of ocean waters. This should result in a slower transfer of ^{14}C to the deep ocean, thereby increasing atmospheric^{14}C. Another climatic factor influencing the rate of exchange of carbon between ocean and the atmosphere is wind speed. Estimations made by Stuiver and Braziunas [178] showed that an observable rise of atmospheric concentrations of radiocarbon during MM can result in the reduction of globally averaged wind speed by two thirds.

There is a clear dependence between the coefficient of attenuation of the rate of ^{14}C generation $Att(T)=\dfrac{\Delta Q/Q}{\Delta N_a/N_a}$, and period T (Fig. **3.5**).

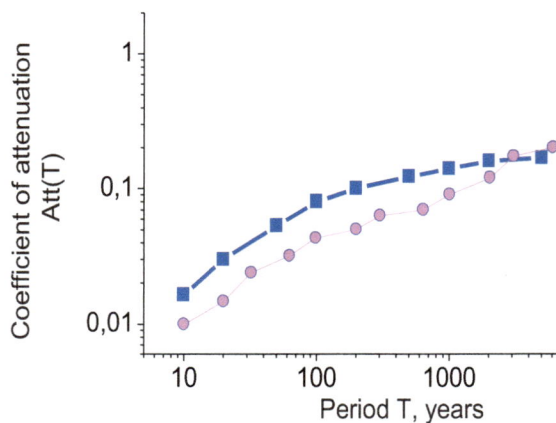

Fig. (3.5). Coefficient of attenuation - relation of the amplitude of oscillation of Q to the amplitude of the corresponding oscillation [^{14}C] in atmosphere. Blue line with squares - coefficient calculated using five-reservoir model with parameters (4.11). Magenta line - coefficient calculated by means of a PANDORA model [179], scanned from [180].

It is evident that the radiocarbon exchange system operates as a low-pass filter (Fig. **3.5**). It weakens the short-term variations of radiocarbon production rate more strongly than long-term variations. For example, the 11-year sinusoidal variation of production rate attenuates by ca 100 times. Sine waves with different periods also have different phase shifts.

The procedure described above can be used for building millennial or longer sunspot reconstructions. The parameters of the radiocarbon-exchange model are determined by fitting a proxy data set to the actual solar activity index over reference time interval (training or calibration) (*e.g.* 1615-1945). A non-linear regression between $\Delta^{14}C$ and instrumental data is carried out and then this regression is extrapolated backward in time and the sunspot numbers of the past can be estimated. Recently, new advances in paleoastrophysics have been achieved when physics-based models have been increasingly applied instead of phenomenological regressions [176, 181].

If we focus research on cosmic-ray induced variations of atmospheric ^{14}C through industrial era (after the end of the 19th century), a correction for anthropogenic effects must be made. In particular, the use of fossil fuels since the second half of the 19th century has led to the release of large amounts of carbon dioxide into the atmosphere. This carbon dioxide has a negligible content of ^{14}C because the source of fossil fuels has almost infinite age. This effect was discovered by H. Suess [182] and named after him. It has lead to a dilution of radiocarbon concentrations in the atmosphere. Changes in atmospheric ^{14}C concentration caused by the input of carbon dioxide produced by combustion of fossil fuels can be described by a simple one-reservoir model. The respective equation has the form [182]:

$$\frac{d(\Delta^{14}C(t))}{dt} + \frac{1}{\tau_r}(\Delta^{14}C(t) - \Delta^{14}C(0)) = -\frac{10^3}{M_a(0)}\frac{dM_f}{dt}, \qquad (3.16)$$

where $\Delta^{14}C$ is in per mille, $^{14}C(0)$ is the concentration of ^{14}C in 1860 (prior to the start of large-scale industrial activities), $\dfrac{dM_f}{dt}$ is the rate of anthropogenic ^{12}C input in Gt $\times$ yr^{-1}, τ_r is the residence time (or time of life of ^{14}C in the atmosphere), and $M_a(0) = 1.0$ Gt is the mass of ^{12}C in atmosphere in 1860. Result of such a correction is appreciable (Fig. **3.6**).

Industrial decrease of atmospheric ^{14}C concentration is about 15 per mille (Fig. **3.6B**). Similar result was obtained by Stuiver and Quay [148], who used a more sophisticated approach.

3.2.2. Short-Term Variations of Radiocarbon

Maunder minimum (AD 1645-1715) is a convenient time interval for studying short-term (period less than 30 years) variations of ^{14}C. This period of exceptionally

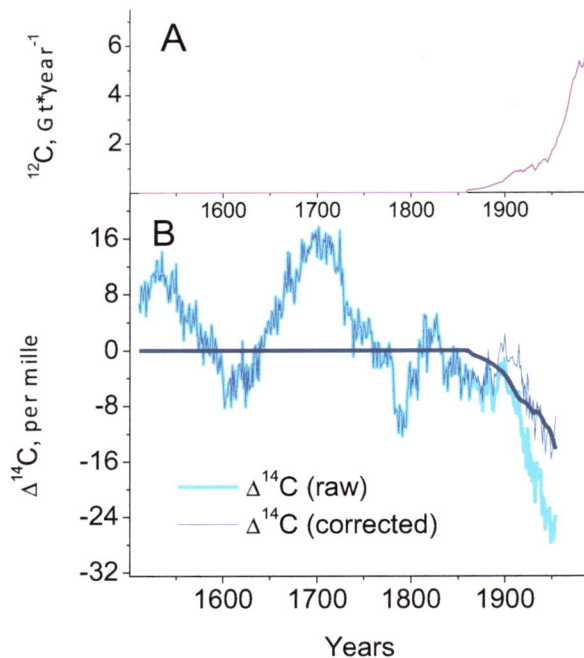

Fig. (3.6). (**A**) human-induced CO_2 emission after Keeling [184] and Joos [185]; (**B**) radiocarbon concentration in tree rings after Stuiver and Braziunas [178], (thick cyan curve); human-induced dilution of ^{14}C concentration calculated with $\tau_r = 6$ years (thick dark blue curve); ^{14}C concentration in tree rings, corrected for Suess-effect (thin blue line).

weak solar activity is of particular interest in scientific papers. For this reason, many radiocarbon series have been obtained for the Maunder minimum period. Some of them are based on measurements of $\Delta^{14}C$ from samples of tree rings (Fig. **3.7**) - they were collected: in Lithuania [186]; in Bashkiria [142, 187]; in the Carpathians and Karelia [188]; in the northeast United States [178]; and in Japan [189]. It is evident (Fig. **3.7**) that any correlation between the specified series measured in different laboratories is practically absent - they are discrepant in both spectral structure and amplitudes of short-term variations.

The possible influence of regional climate on ^{14}C concentration in tree-rings was suggested by [190, 191]. In that case, a distinction between different data sets could arise as a result of particular climatic properties of the regions from which the samples were taken. The hypothesis apparently is out of a standard assumption about the well-mixed tropospheric radiocarbon. On the other hand, Damon *et al.* [192] found an appreciable correlation between $\Delta^{14}C$ in tree-rings collected in the

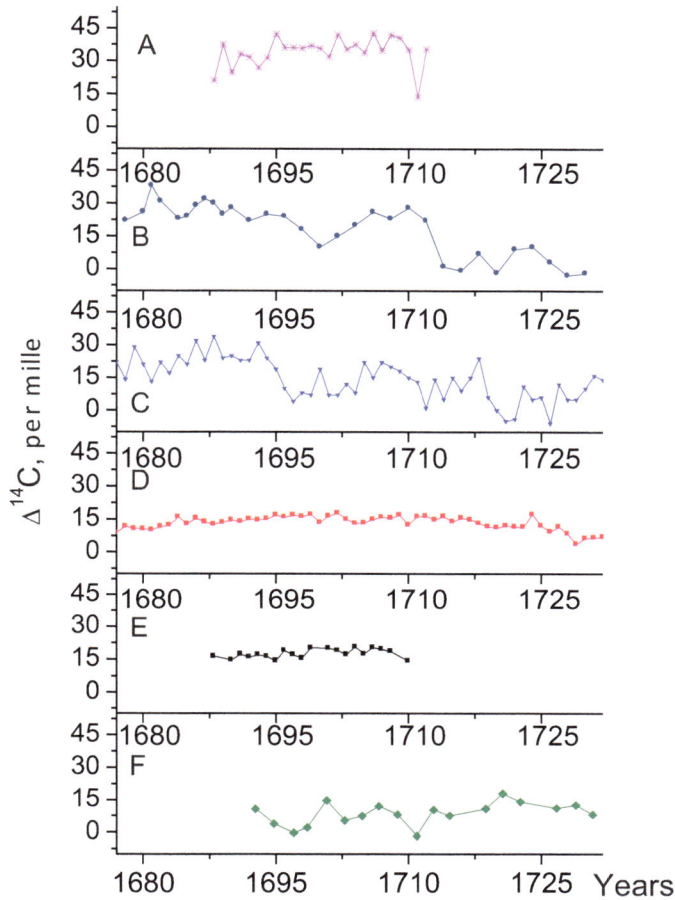

Fig. (3.7). Concentration of ^{14}C in tree-rings measured in: (**A**) Ioffe PhTI (samples from Lithuania [186]); (**B**) Ioffe PhTI (samples from Bashkiria [187]; (**C**) Tbilisi University (samples from Carpathians and Karelia [188]); (**D**) Washington University (samples from north-west USA [178]); (**E**) University of Arizona (samples from Bashkiria [187]); (**F**) University of Nagoya (samples from Japan [186]).

delta of Mackenzie River with the mean May-August local temperature. In addition McCormac *et al*. [193] arrived at a conclusion that ^{14}C concentrations have substantial fluctuations on a regional scale. However, even if the contribution of a local climate to the concentration of ^{14}C in rings of trees is really significant, an account of the unlikely regional environmental conditions can explain completely the differences between experimental series measured in different laboratories. The matter is that considerable divergences are observed in series obtained from geographically-close areas, even when identical samples are used (Fig. **3.8**).

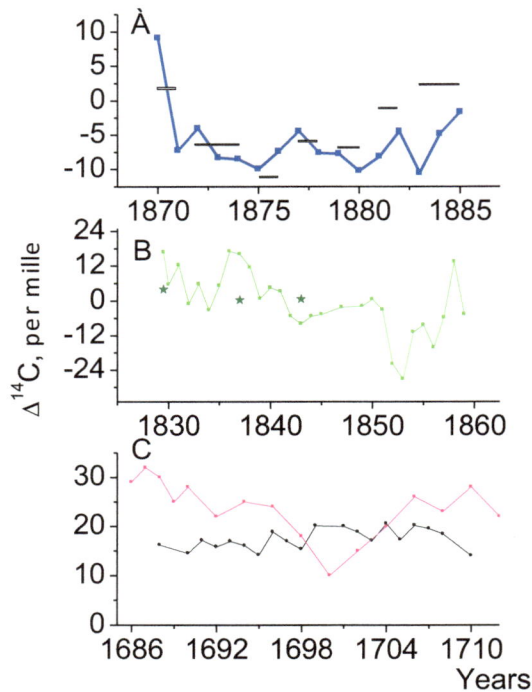

Fig. (3.8). Concentration of ^{14}C measured in: (**A**) rings of trees growing in the delta of Mackenzie. Blue line with squares - [192], thin rectangles - [194]. Both curves scanned from [192] and digitized. (**B**) rings of trees growing in England. Green line with squares - [195], stars - [148]. Both curves scanned from [148]. and digitized. (**C**) rings of trees growing in Bashkiria (USSR). Pink line - [142]; black line - [187].

Differences between the experimental series obtained using practically identical samples are significant (Fig. **3.8**). This disagreement is seemingly very difficult to explain. One can speculate that some local noise may originate in a laboratory. Some very large (1-2%) amplitudes of Δ^{14}C fluctuations in series after [139, 194, 195] are also difficult to explain. In any case, the possibility looks doubtful for using radiocarbon data for the detection and reconstruction of short-term (11-year and 22-year) variations of solar activity and GCR intensity. It is most likely that the high-frequency component (fluctuation with the periods less than 30 years) of radiocarbon series does not carry considerable astrophysical information.

3.2.3. Mid- and Long-Term Variations of Radiocarbon

In contrast to previous discussions of short-term variations, here, we show that radiocarbon records make it possible to reconstruct multi-decadal and century-long variations of sunspot activity quite accurately at least over the last 4 centuries (Fig. **3.9**). That illustrates the astrophysical significance of the mid-term (30 years-few centuries) variation of ^{14}C in tree rings. A reasonable agreement between different experimental series confirms this conclusion.

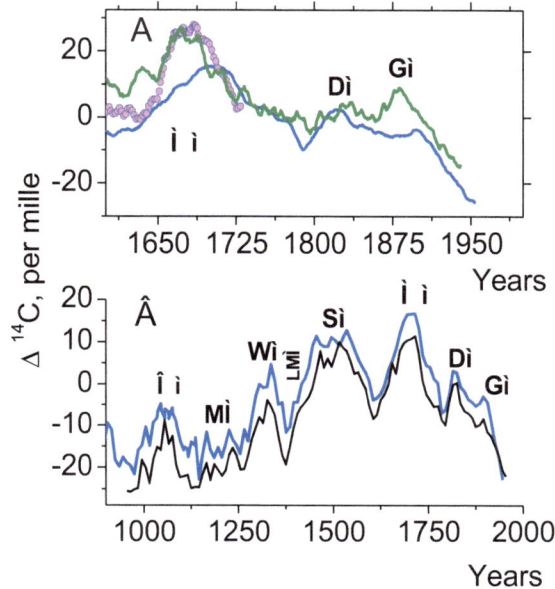

Fig. (3.9). Concentration of ^{14}C in tree-rings measured in: (**A**) Tbilisi State University (samples from Karelia and Carpathians [188]) - green line; Ioffe PhTI (samples from Bashkiria [142]) - magenta circles; Washington University (samples from north-west USA [178]) - blue line. All the curves are smoothed by 13 years. (**B**) mean decadal concentration of ^{14}C measured in trees from Northern Hemisphere - blue line [171] and Southern Hemisphere - black line [193]. Abbreviations are: Om - Oort minimum, MM - Medieval maximum, Wm - Wolf minimum, LMM - late Medieval maximum, Sm - Spoerer minimum, Mm - Maunder minimum, Dm - Dalton minimum, Gm - Gnevyshev minimum.

Three annual radiocarbon records measured in Ioffe PhTI [142], Tbilisi State University [188], and at the University of Washington [178] and averaged over 13 years demonstrate that, in spite of some differences, correlation between the data is appreciable - the coefficient of correlation reaches $r_l = 0.7$ (Fig. **3.9A**). The mean decadal (10-year average) value of $\Delta^{14}C$ in tree rings from the Northern Hemisphere (samples collected from Germany, Ireland, and in the northwest United States) and the Southern Hemisphere (samples collected from Southern Africa, New Zealand and Chile) were measured by [171] and [193] correspondingly (Fig. **3.9A**). It is evident that century-scale fluctuations of the concentration of ^{14}C in the atmospheres of the two hemispheres occur almost synchronously (Fig. **3.9B**). The global minima and maxima of solar activity are distinctly manifested in the radiocarbon data for both hemispheres. A small shift between curves is most likely caused by geographical differences between the hemispheres. Similarity in century scale variations of Δ^{14}C in rings of trees growing in different parts of the Earth clearly points to common origins and extraterrestrial sources. It is evident that variation of GCR intensity is the most probable common factor. Thus, experimental data on medium-frequency radiocarbon variability corroborate their importance for solar paleoastrophysics.

Two radiocarbon paleoreconstructions of more than 10 000 years in length [176, 196] and a reconstruction covering ca 5000 years [194] were obtained from the same IntCal98 series [168] but using different reconstruction procedures (Fig. **3.10**). It is seems that while an agreement between the three series over the multi-decadal to centennial scale is obvious, the long-term (many centuries to millennia)

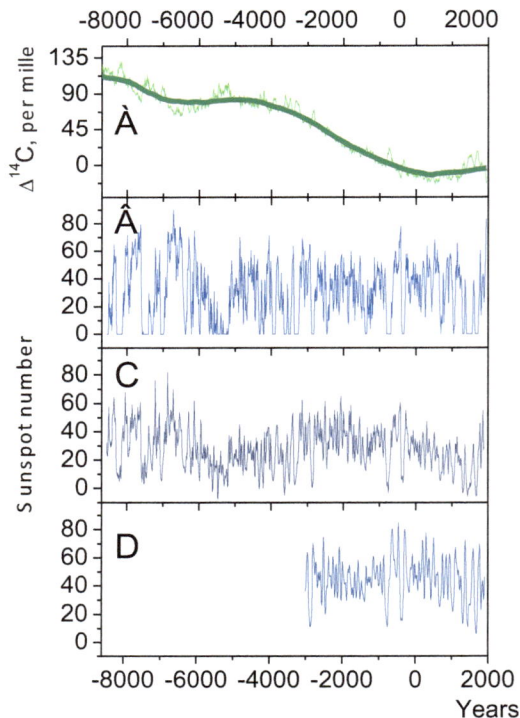

Fig. (3.10). (A) radiocarbon series IntCal98 [171]. Thick dark green line - long-term trend. **(B)** sunspot numbers reconstructed in [198]; **(C)** in [176]; **(D)** in [197].

variations differ. One possible reason for this dissimilarity could be the difference in trend removal. Secular (millennial and longer) variation of $\Delta^{14}C$ usually is attributed to changes in geomagnetic dipole which affect the geomagnetic shielding of the cosmic ray flux. The methods applied for detrending could, however, be different. For example, Ogurtsov [198] subtracted the long-term trend by pure mathematical procedure (removing of 2000-year average). Solanki *et al.* [176] removed the trend using experimental data on the evolution of the geomagnetic dipole moment.

The data series on variations of the Earth's dipole moment, reconstructed from sedimentary records during the last 10,000 years show that the divergence between the available paleomagnetic records could reach 15-20% during the last 8000 years and more than 30% in the more remote past (Fig. **3.11**). Thus, a reliable reconstruction of the Earth's magnetic dipole field over secular-to-millennial timescales is still problematic. It is therefore not currently possible to accurately separate geomagnetic from solar modulation [199].

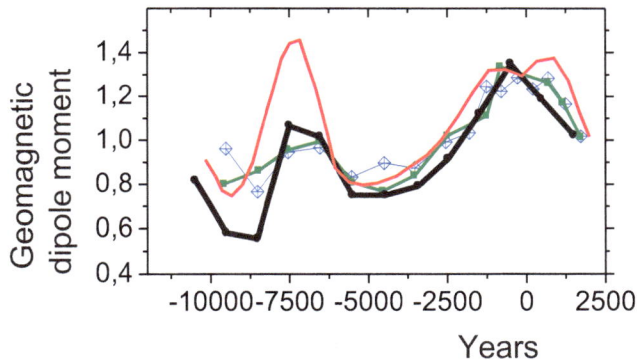

Fig. (3.11). Geomagnetic field of the Earth relative to the modern value. Red line - according to [199], black line - according to [200, 201], green line with squares - according to [202], blue line with rhombs - according to [203]. The data scanned from the work [204] and digitized.

Insufficiency in our knowledge about the past geomagnetic field intensity is an important source of uncertainty in long-term variations of the reconstructed sunspot numbers. In addition, the removal of multi-millennial trends by means of quite arbitrary mathematical procedures (*e.g.* 2000-year average by Ogurtsov [198], 5th order polynomial by Sonett [205], or local polynomial approximation of 2nd order [206]) brings ambiguity to radiocarbon reconstructions. Moreover, deficiencies in current understanding of the processes of ocean ventilation can also contribute to the uncertainty of reconstruction. The oceans are the largest radiocarbon reservoirs, including more than 90% of the terrestrial ^{14}C. They absorb ^{14}C produced in the atmosphere and reintroduce it into the atmosphere with some delay. The delay is determined by a lifetime of radiocarbon in the deep ocean, which is dependent on processes of oceanic dynamics. Some estimations of timescales of deep ocean circulation range many centuries [207, 208] to a few millennia [209]. This means that the assumption of a well-mixed deep ocean used in many-reservoir models is not fully substantiated for processes which last for millennia. The hypothesis of quasi-homogeneous deep ocean appears to be strongly simplified in relation to the variations of Q, with periods comparable with a timescale of deep ocean circulation. In that case, a consideration of the slow distribution of ^{14}C in a diffusive deep ocean medium accounting for upwelling - a wind-driven vertical transport of deeper water to shallow levels - could be a useful improvement of the model. Main zones of upwelling in the world's oceans occur in California, Peru, Canarias, and Benguel currents. In the framework of this approach, the distribution of ^{14}C in deep ocean could be described by the equation [210]:

$$\frac{\partial C_d(z,t)}{\partial t} - K_z(z)\ \frac{\partial^2 C_d(z,t)}{\partial z^2} - \frac{\partial K_z(z)}{\partial z}\ \frac{\partial C_d(z,t)}{\partial z} + w_z(z)\ \frac{\partial C_d(z,t)}{\partial z} + \lambda \cdot C_d(z,t) = 0, \text{ (3.17)}$$

where $C_d(z,t)$ is the concentration of ^{14}C in the deep ocean in at×cm^{-3}; z is depth (vertical distance from the bottom surface of the mixed layer) in cm; $K_z(z)$ is a coefficient of vertical diffusion in cm^2× s^{-1}; and $w_z(z)$ is an upwelling velocity in cm×s^{-1}. The equation (3.17) is written with the assumption that the downward transport of radiocarbon is provided by turbulent diffusion (which operates towards the direction of [^{14}C] decrease), while upward transport is provided by upwelling. Indeed, over a major part of the World Ocean, the average speed of the vertical movement of water is directed upwards. It follows from: (a) condition of balance of water masses - upwelling compensates powerful downward water fluxes in areas of deep waters formation (Greenland, Norwegian, and Labrador seas in the Northern Hemisphere; Weddell and Ross seas in the Southern Hemisphere) and (b) conditions of thermal balance - that is, the transfer of heat by upwelling flux compensates the diffusion heat flux from a warm surface of an ocean to its cold depths. The account of thermal balance makes it possible to link upwelling velocity with the vertical diffusion:

$$w_z\ T\ S = K_z\ \frac{dT}{dz}\ S \rightarrow w_z = \frac{K_z}{T}\ \frac{dT}{dz}, \qquad \text{(3.18)}$$

where S is the ocean surface. If we take $T = 3.6°C$ - mean temperature of the World Ocean [211], $\dfrac{dT}{dz} = \dfrac{(T_s - T_d)}{H_0}$ - mean vertical gradient of temperature; $T_s = 17°C$ - mean temperature of the mixed layer; $T_d = 2°C$ mean temperature of the deep ocean; and $H_0 = 5000$ m - mean depth of the ocean; we obtain $w_z = 8.3 \times 10^{-6}\ cm^{-1}\ K_z$. Clearly, the coefficient of vertical diffusion K_z is the most important parameter for the description of ^{14}C distribution in the deep ocean. However, available data on turbulent diffusion in the ocean are not in good agreement in terms of estimations of global and regional coefficients of the vertical diffusion, achieved by studying the microstructures of the deep layers of ocean, estimations of balance of water masses, and an analysis of the distribution of different tracers (Table **3.3**).

Table 3.3. Experimental and theoretical estimations of the coefficient of vertical diffusion in different parts of the World Ocean.

Source	Geographic Region	Way of Estimation	Depth	$(cm^2 \times s^{-1})$
Stuiver and Quay [148]	World Ocean	distribution of ^{14}C, Suess-effect	all the depth range	3.0
Kunze and Sanford [212]	Sargasso sea	measurement of the depth microstructure	all the depth range	0.1
Bryan and Lewis [213]	World Ocean	water mass balance	surface layer	0.3
Tsujino *et al.* [214]	World Ocean	model	surface layer	0.1
King and Deval [215]	eastern part of Pacific Ocean, tropics	nitrate uptake by phytoplankton	surface layer and upper thermocline	0.05-1.1
Stuiver [216]	Atlantic Ocean	distribution of ^{14}C	surface layer and thermocline	4.0
Ledwell *et al.* [217]	North Atlantic ($\varphi = 26°$ N, $\lambda = 29°$ E)	distribution of SF_6	thermocline	0.11
Schmitt *et al.* [218]	North Atlantic ($\varphi = 10$-$17°$ N, $\lambda = 50$-$70°$ E)	distribution of SF_6 and salinity	thermocline	0.8-0.9
Munk [219]	Pacific Ocean	distribution of ^{14}C	>1000 m	1.3
Roemmich *et al.* [220]	Samoa passage	heat balance	bottom layer	500
Led well *et al.* [221]	South Atlantic, Brazil Basin	distribution of SF_6	bottom layer	10
Tsujino *et al.* [214]	World Ocean	model	bottom layer	3.0

Uncertainty in available estimations of the globally-averaged values of K_z (Table **3.3**) reaches a factor of 2-3 or even more. In addition, K_z has appreciable geographic variability - its value at different parts of the World Ocean can differ in tens of times. Furthermore, information about the depth profile of diffusion coefficient is not exact. The situation with upwelling velocity is similar. Thus current knowledge about long-term processes of ^{14}C distribution in deep ocean layers is insufficient. For this reason it is not easy to decide:

(a)   whether the secular-to-millennial changes in ^{14}C production rate were caused by changes in solar activity or by a varying geomagnetic field.

(b)   which part of the secular-to-millennial variations of atmospheric $\Delta^{14}C$ value was caused by changes in the production rate and which part by variations in oceanic dynamics.

Presently, the study of variations of the Sun's magnetic activity over time scales of an order of several thousands of years or more and using radiocarbon data is very demanding. Medium-term (multidecadal-to-centennial scale) variations of $\Delta^{14}C$ in the atmosphere are the most valuable from the point of view of solar paleoastrophysics.

3.3. Variations of ^{10}Be in Polar Ice and their Paleoastrophysical Implementations

The peak to trough amplitude of 11-year variation of ^{10}Be production rate is ca 25%; or close to that of radiocarbon. A very direct recording of beryllium production rate changes this variation attenuates only by 20% [222]. This makes it possible to trace 11-year solar variations using ^{10}Be records.

3.3.1. Short-Term Variations of ^{10}Be

Beer *et al.* [116] have shown that the concentration of ^{10}Be in the Greenland ice has the distinct eleven-year periodicity correlating with the corresponding cycle in Wolf numbers (Fig. **3.12**). Estimations of [116] showed that during AD 1783-1985, the coefficient of annual correlation between Wolf number and beryllium concentration is 0.45 with a one year lag corresponding to 7-14-year periods.

Fligge *et al.* [223] used beryllium record for studying variations of the length of a solar cycle of Schwabe throughout more than five centuries. Steig *et al.* [224] successfully used the 11-year cycle of ^{10}Be concentration for dating the ice core retrieved from east Antarctica over large depths, where it was impossible to distinguish individual annual ice layers.

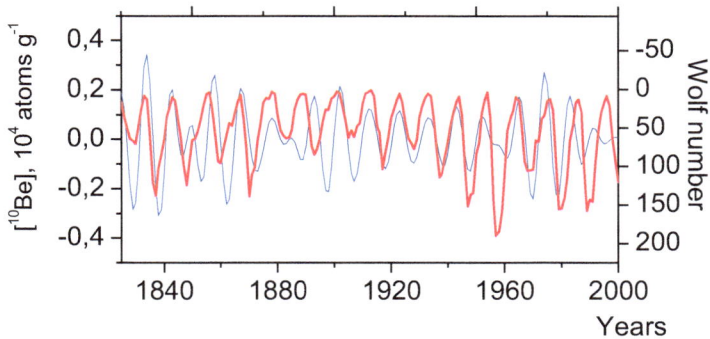

Fig. (3.12). Concentration of ^{10}Be in South Greenland ice (Dye-3 station, blue line) [116, 165] and Wolf number (red line), filtered over the frequency band.

It is obvious that the high-frequency (periods less than 30 years) component of beryllium variability contains valuable information on past changes of the Sun's activity and GCR intensity. On the other hand, manifestations of a solar cycle of Schwabe in beryllium records sometimes look a bit unusual. For example, the amplitude of a quasi-eleven-year variation of ^{10}Be concentration through the Maunder Minimum was not smaller than during epochs of normal solar activity [225]. Usoskin *et al.* [226] assumed that just during the periods of deep minima of the Sun's activity, the beryllium concentration in ice can be substantially distorted by climatic effects. Beer *et al.* [116] also have noted that short-term meteorological fluctuations can affect ^{10}Be concentration.

In any case, beryllium records measured in polar ice can serve as indicators of short-term changes of solar activity in the past. That testifies to the paleoastrophysical significance of short-term variations of ^{10}Be.

3.3.2. Mid- and Long-Term Variations of ^{10}Be

Two annually resolved beryllium time series

(a) ^{10}Be record after [117] measured from an ice core from Dye-3 site in South Greenland (φ=65.18^0 N, λ=43.83^0 W, altitude 2870 m), which covers the time interval AD 1424-1985) (Fig. **3.13A**); and

(b)   A new ^{10}Be series covering the time interval AD 1389-1994, which has been measured from an ice core retrieved from NGRIP site in Central Greenland (φ=75.10° N, λ=42.32° W, 2917 m) [117] (Fig. **3.13B**), both

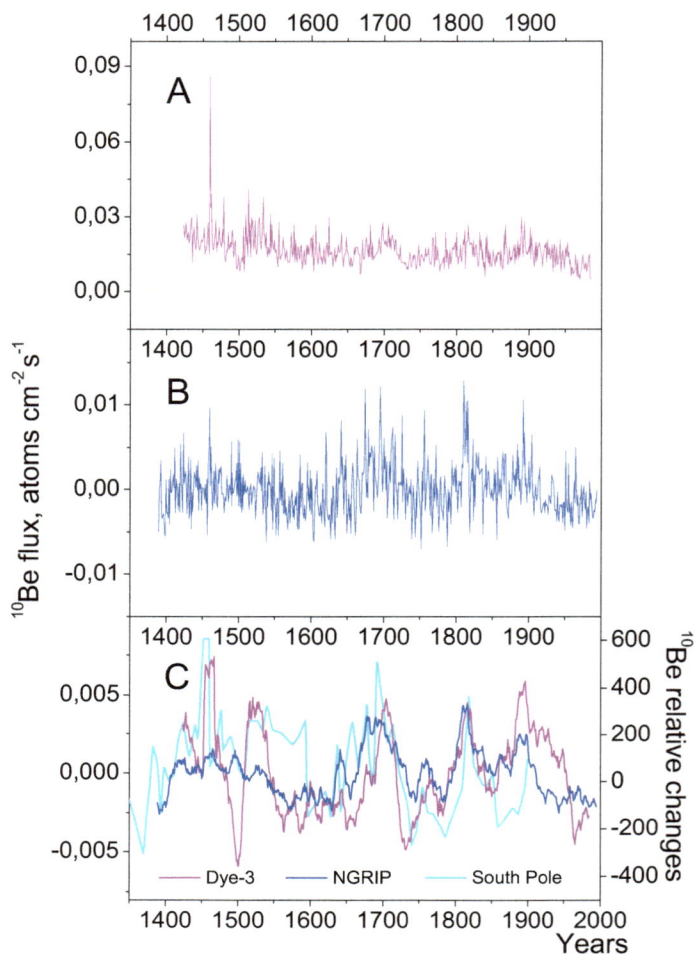

Fig. (3.13). Beryllium flux calculated: (**A**) from Dye-3 core (South Greenland, [116, 165]); (**B**) from NGRIP core (Central Greenland, [165]); (**C**) Dye-3 and NGRIP series, averaged over 15 years (left scale). South Greenland - magenta line, Central Greenland - blue line. Cyan line - concentration of ^{10}Be at South Pole (right scale, [162]).

smoothed by a 15 year moving average, are rather well synchronized after the mid-16th century (Fig. **3.13C**). The coefficient of correlation calculated over AD 1550-1985 is 0.52. Thus the ^{10}Be records obtained from different geographic locations have apparent similarity over time scales longer than ca 20 years. That testifies for the presence of a source of common origin for such variability; one which is not linked to climatic changes. Obviously, the changing fluxes of high-energy cosmic rays, effectively modulated by the Sun's activity, could be such an origin. The complexity of the ice flow regime could result in a weaker relationship between ^{10}Be concentration in ice with the cosmic ray intensity and solar activity in the earlier part of the records [117]. Another series of ^{10}Be concentrations has been measured from a core retrieved from the South Pole [162] (Fig. **3.13C**). The Antarctic record also has evident common features with the Greenland series. Thus, ^{10}Be concentration in ice looks like quite reliable proxy for mid- and long-term changes of the Sun's activity. Sunspot numbers, reconstructed for the last 400 years from ^{10}Be by Usoskin *et al.* [181] (Fig. **3.14**), show that the agreement between the reconstructed sunspot number and the measured sunspot number is quite good at least after the end of Maunder Minimum. The correlation coefficient is $r_l = 0.72$ over the entire interval of reconstruction (1610-1985). This is taken as evidence that multi-decadal and centennial variations of beryllium concentration really reflect the corresponding variations of solar activity.

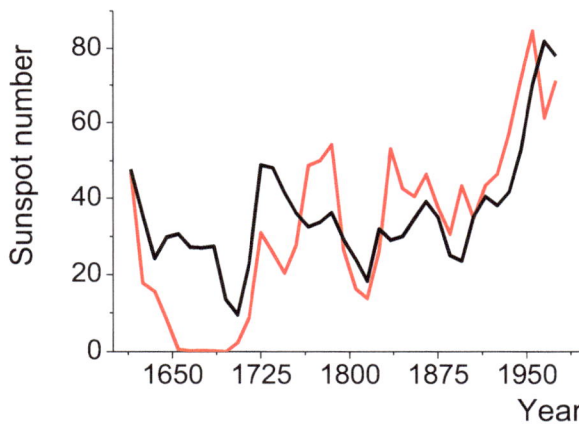

Fig. (3.14). Actual (red line) and reconstructed (black line) group sunspot number smoothed and interpolated by decades. Reconstructed sunspot number was electronically scanned from [181] and digitized.

Analysis performed by Bard *et al.* [162] confirmed connection between century-type variations in sunspot number and ^{10}Be concentration in ice.

Secular-to-millennial variations of beryllium concentration in ice have also been studied in a number of works [227, 228]. Ogurtsov [228] studied the concentrations of ^{10}Be and ions of Na^+, Ca_2^+ in the ice core retrieved from Central Greenland (72.4° N, 38.3° W), which was measured by specialists of the GISP2 collaboration [229-232]. The temporal series obtained within the GISP2 project cover a substantial part of the Pleistocene period and the ^{10}Be concentration was measured for the past 40 000 years (with lacunas). Fig. (**3.15A-C**) depicts the data on the concentration of the cosmogenic isotope [10]Be [229-232] and the concentration of Na[+] and Ca_2^+ ions [231, 232] in central Greenland, which were obtained by the GISP2 collaboration. The geomagnetic dipole moment, estimated by Guydo and Valet [233] and Knudsen *et al.* [234], is shown in Fig. (**3.15E**). All time series depicted cover the past 40 000 years.

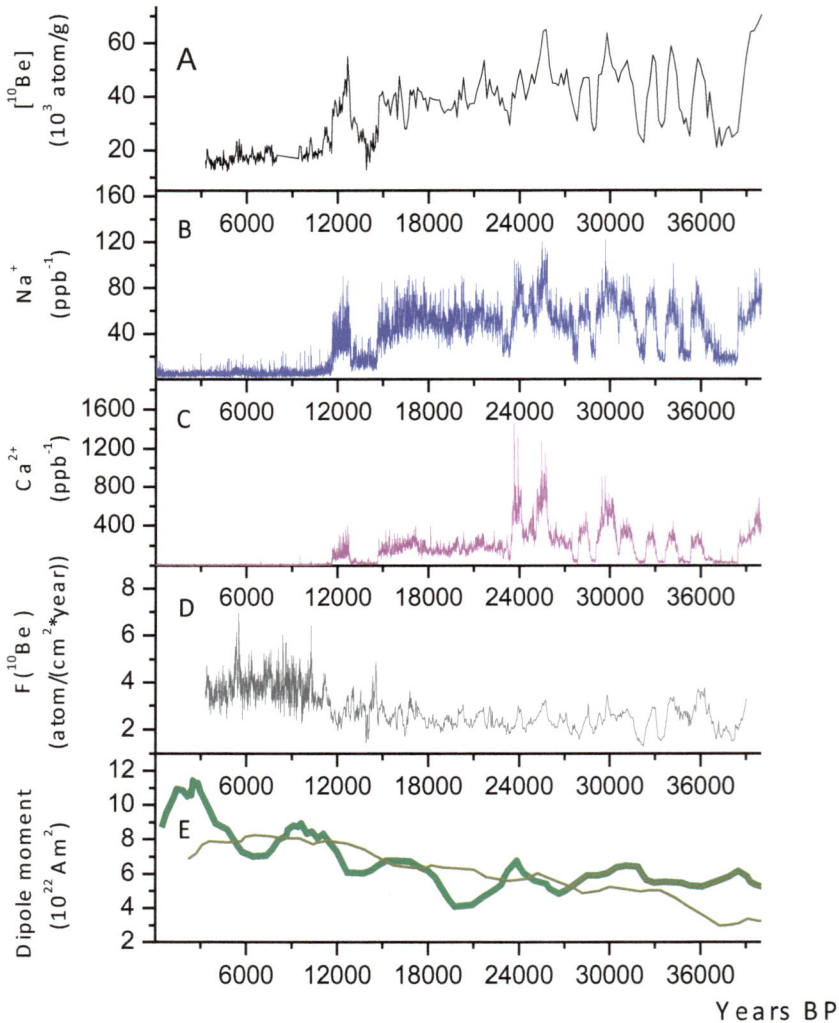

Fig. (3.15). (**A**) [10]Be concentration in ice of Central Greenland; (**B**) Na[+] concentration in ice of central Greenland; (**C**) Ca_2^+ concentration in ice of Central Greenland; (**D**) [10]Be flux in Central Greenland; (**E**) geomagnetic dipole moment according to [233] (thin line) and [234] (thick line).

Na^+ and Ca_2^+ ions enter ice as part of marine and mineral dust aerosols. Therefore, their concentration in ice (*e.g.* from Greenland) substantially depends on the directions of winds bringing air masses to the ice sheet. As a result, the concentration of these ions can be assessed as a proxy for atmospheric circulation in the past [231], indicating that the variations of $[^{10}Be]$ after 12000 BP are very similar to the corresponding variations of $[Na^+]$ and $[Ca_2^+]$ (Fig. **3.15A-C**). Their correlation coefficients, calculated for the time series interpolated by decades during 12008-39008 BP, are 0.78 and 0.75 (ca 50% of common variance), respectively, which demonstrates the strong climatic dependence of the beryllium concentration in ice during this period. The flux of $F(^{10}Be)$ (Fig. **3.15D**) is calculated using formula (3.11), with ice density of 1.0.

According to Ogurtsov [228], the association between $F(^{10}Be)$ and climatic paleoindicators appears to be substantially weaker, especially prior to 23 500 BP. A link between millennial-scale variations of beryllium flux and geomagnetic field intensity is also weak over 12008-23498 BP. The substantial weakening of meridional circulation about 12 000 years ago caused a slackening of the inflow of air masses from the middle latitudes and dominance of high-latitude air masses carrying tropospheric ^{10}Be of local origin, which is insensitive to variations in geomagnetic field strength [228]. Thus, between 12000-23500 BP, the ^{10}Be flux appears to depend weakly on both the climate and the geomagnetic moment (Fig. **3.15B-E**). That is why Ogurtsov [228] attributed the cyclicities with periods of 180-300, ~630 and ~1350 years, existing during 12000-23500 BP to variations in solar activity.

More evidence of the existence of a solar cycle within a period close to 1500 years was presented by Bond *et al.* [235], who examined both ^{10}Be and ^{14}C records. These results testify that the concentration of cosmogenic beryllium in polar ice contains information about past changes of solar activity over time scales from decades to millennia.

An analysis of the available information about the abundance of cosmogenic isotopes in natural archives shows that the data concerning beryllium is more informative. They allow researchers to study not only multi-decadal to centennial scales of solar variability but also the fluctuations of activity of the Sun over short (less than 30 years) and probably on even longer (thousand years or more) time scales. It should be noted that at sites where stratospheric beryllium is deposited, the concentration of ^{10}Be is effectively influenced by geomagnetic fields which multi-millennial variations are uncertain.

3.4. Nitrates - Mechanisms of Formation and Fixation in Terrestrial Archives

The abundance of nitrate ions (NO_3^-) in the polar ice of Greenland and Antarctica has been investigated for many years [125, 236-240]. The properties of nitrate concentration in ice are connected with the mechanism of generation (Fig. **3.16**). According to Logan [241], Legrand and Kirschner [238], and Mayewski *et al.* [242] nitrate "precursors" - the various chemical species of odd nitrogen NO_x (N, NO, NO_2) and NO_y (NO_x, N_2O_5, HN_2O_5, HO_2NO_2, $BrONO_2$, $ClONO_2$,) - are produced at different altitudes of the atmosphere (Table **3.4**):

(a)   In the troposphere (owing to industrial emissions, biomass combustion, soil denitrification, lightning, and the influence of GCR).

(b)   In the stratosphere and higher altitudes (due to biogenic N_2O oxidation $N_2O+O(^1D) \rightarrow 2NO$, GCR, SCR, solar UV radiation, and relativistic electron precipitation).

Table 3.4. Sources of NO molecules in the Earth's atmosphere. Data were taken from [101, 243, 244].

Source	Domain of Generation	NO (Molecules/Year)
$N_2O+O(^1D)$	Stratosphere	4.5×10^{34}
GCR	Upper troposphere + Stratosphere	$2.7\text{-}3.7 \times 10^{33}$
REP	Lower thermosphere + Mesosphere	$1.4 \times 10^{33}\text{ -}1.4 \times 10^{34}$
Meteors	Mesosphere	6.3×10^{32}
Lighting	Troposphere	1.6×10^{36}
SPE	Upper stratosphere + Mesosphere	up to $6\text{-}8 \times 10^{33}$

Nitrate produced by cosmic radiation are important for the solar paleoastrophysics. Fluxes of energetic solar and galactic cosmic ray particles as well as precipitations of relativistic electrons from the radiation belts generate a lot of secondary free electrons with energies of hundreds of eV, which effectively interact with molecules of atmospheric gases causing their ionization, dissociation, and excitation:

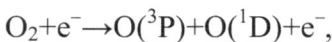

$$O_2+e^- \rightarrow O(^3P)+O(^1D)+e^-,$$

$$N_2+e^- \rightarrow N^+ +N(^4S,^2D)+2e^-, \tag{3.19}$$

$$N_2+e^- \rightarrow N(^4S) +N(^4S,^2D,^2P)+ e^-.$$

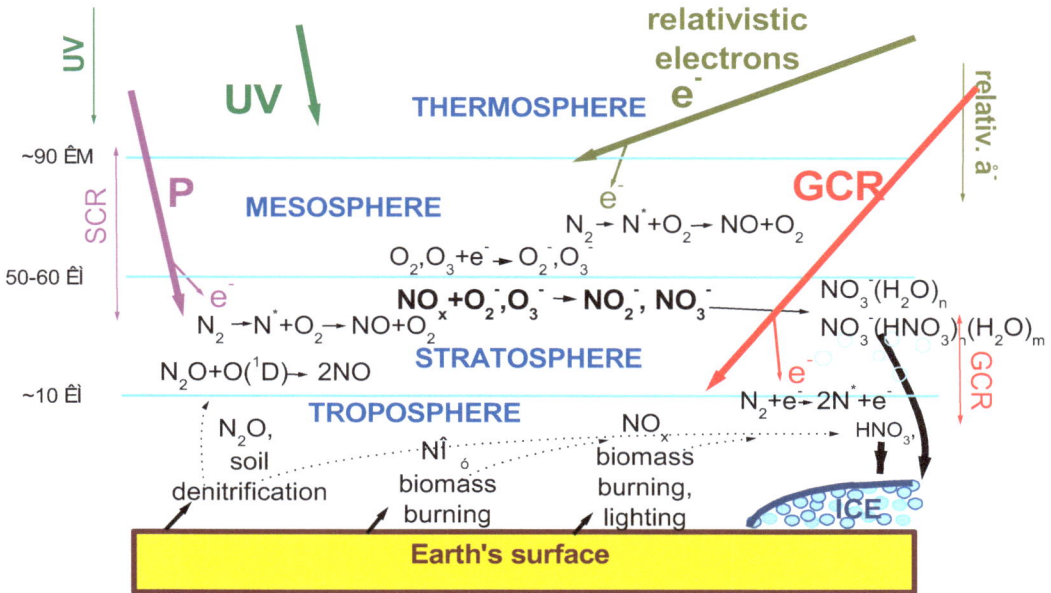

Fig. (3.16). Generation of a nitrate record in polar ice.

According to [101], a strong proton event of 19-27 October 1989 produced 6.6×10^{33} atoms and molecules of NO_y and another proton event of 2-10 August 1972 3.6×10^{33} atoms and molecules NO_y. Different NO_y species, the ions O_2^-, N^+, O^+, N_2^+, O_2^+, excited atoms of oxygen and nitrogen, which are involved in a chain of photochemical reactions resulting in the production of nitrogen oxide NO. The lifetime of ions in the stratosphere is usually in the order of several minutes. Jackman *et al.* [245] reported that each pair of ions produced by 20-200 MeV protons creates ca 1.5 NO molecules. For lower energy (tens of keV) among auroral electrons the corresponding number is 2.0-2.68 [246]. Therefore, an association appears between the atmospheric ionization and the production of nitrogen oxides. NO oxidizes to NO_2, which in turn serves as a source for nitrate ions:

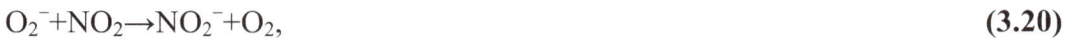

$$O_2^- + NO_2 \rightarrow NO_2^- + O_2, \tag{3.20}$$

$$NO_2^- + O_3 \rightarrow NO_3^- + O_2.$$

Reactions with other NO_y species can also play some role in nitrate ion production. For example, heterogeneous reactions on polar stratospheric cloud particles [247]:

$ClONO_2+HCl \rightarrow Cl_2+HNO_3,$

$N_2O_5+HCl \rightarrow ClNO_2+HNO_3,$ (3.21)

$HNO_3+OH^- \rightarrow NO_3^- +H_2O.$

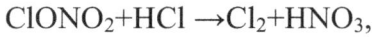

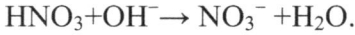

The key mechanism of the stratospheric NO_y loss is [245]:

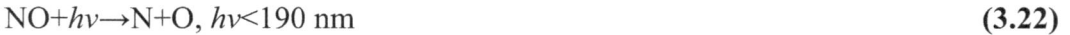

$NO+hv \rightarrow N+O,$ $hv<190$ nm   (3.22)

$N+NO \rightarrow N_2+O.$

Thus, odd nitrogen survives better during the polar night in the absence of sunlight. That is why the NO_y produced by precipitating particles mainly at the polar caps at high latitudes can be transported to the surface. NO_3^- ions can produce clusters, particularly with water molecules:

$NO_3^- HNO_3,$ NO_2^- $(HNO_2)H_2O,$ $NO_3^-(H_2O)_n,$ n=2-5   (3.23)

The hydration of nitrate ions occurs quickly enough because concentration of molecules H_2O in the stratosphere is high enough, 4-6 per million molecules of air. Since the concentration of molecules of air at an altitude of 25 km (P=30 HPa, T=230° K) is about 10^{18} cm^{-3}, the corresponding value of [H_2O] is close to 5×10^{12} cm^{-3}. Thus, according to Webber [248], the frequency of collisions between ions and water molecules is:

$$v = 4.4 \cdot 10^{-15}(m^3 \times s^{-1}) \left(\frac{T}{300} \right) n_{H_2O} = 1.6 \times 10^4 \ s^{-1}$$ (3.24)

These cluster ions have a long lifetime (up to 10^3-10^4 seconds) and thus can be absorbed by aerosol particles. Afterwards they precipitate on the terrestrial surface both by gravitation sedimentation and descending air streams. The downward streams facilitating delivery of ions NO_3^- to the terrestrial surface are distinctly expressed in high-latitude areas of the Earth, whose stratosphere and upper troposphere (altitudes 10-20 km) is enveloped in a circumpolar vortex during winter. The border of the vortex lies approximately at latitude 70°. Over the southern circumpolar vortex, the vertical speed of descending streams of air near the tropopause is 0.03-0.1 km/day [249, 250]. Another source of nitrate ions in ice is nitric acid HNO_3, which is produced in the troposphere due to a reaction of NO_x with water. Nitrate is deposited from the atmosphere to snow through wet and dry deposition. For dry deposition, the nitrate flux should be constant over a wide range of ice accumulation values. Dry deposition dominates over sites with relatively low accumulation rates (for example, sites with less than 100 mm of water equivalent per year in Antarctica) [130]. At sites with even lower accumulation rates (those with less than 50 mm of water equivalent per year in Antarctica), post-depositional effects may alter the nitrate content in ice [130]. The re-evaporation of nitric acid and photolysis of NO_3 in the upper snow layer are the likely causes of post-depositional changes [251]. Despite that, a major part of odd nitrogen is produced at high latitudes; Randall *et al.* [252] reported that solar storms in 2003-2004 caused an increase of stratospheric NO_x and a decrease of O_3 over latitudes up to 40° N.

Because of the large number of NO_x sources and their considerable geographical variability, the choice of a site for the drilling of a core to be used in paleoastrophysical research poses a serious problem. The basic guideline strategy for the selection of a point for drilling is to search for areas over which: (a) atmosphere concentration of NO_x produced by cosmic rays is maximal and (b) the intensity of local meteorological fluctuations is minimal. Geomagnetic rigidity cut-off is low in the high-latitude region of the Earth and thus penetration of cosmic ray particles into atmosphere is facilitated. This area evidently satisfies the first condition. Moreover, the absence of light during long polar nights promotes the accumulation of odd nitrogen in the atmosphere. Local meteorological instabilities are connected mainly to cyclonic activity. Anticyclones are much less mobile - they can stay at the same place over several days. Pressure gradients associated with anticyclones are less than cyclonic ones, and calm is often observed in the anticyclone centre. Central parts of large ice sheets - Greenland and Antarctica - are the regions which are the most free from cyclonic activity. Over these regions, a near-surface radiation balance (accounting of the incoming and outgoing components of radiation) is negative because of high albedo. Over central Greenland, the radiating balance is -7 $W \times m^{-2}$ [253] and over internal areas of Antarctica it is -15-20 $W \times m^{-2}$ [254]. This means that vast areas of continental ice caps do not heat surface layers of air (as all other parts of the planet do) but cool them by working as refrigerators. Since a powerful ascending air motion is one of the key factors providing existence of a cyclone, the cooling of the surface air layer, which makes it denser and heavier, evidently hampers cyclone generation. It is easy to estimate the influence of the Greenland ice sheet on cyclonic activity. The total area of the Greenland glacier is $S_{Greenl} = 1.7 \times 10^6$ km^2. Considering $\Delta W_{Greenl} = -7$ $W \times m^{-2}$, a cooling capacity of the ice sheet could be estimated as:

$$\Delta E_{\text{Greenl}} = S_{\text{Greenl}} \times \Delta W_{\text{Greenl}} = 1.7 \times 10^{12} \text{ m}^2 \cdot 24 \cdot 3600 \text{ s} \cdot 7 \times 10^7 \text{erg} \times \text{m}^{-2} \times \text{s}^{-1}$$
$$= 1.0 \times 10^{25} \text{ erg/day}. \tag{3.25}$$

The power of an average cyclone is ca 10^{25} erg/day and the characteristic area ($\cong 2 \times 10^6$ km^2) is close to the area of the Greenland ice sheet. Thus the Greenland glacier can provide the dissipation of a cyclone's kinetic energy with average sizes and power [255]. For this reason, an average cyclone settled down over Greenland, for example, will start to collapse and will be quickly destroyed. Only powerful cyclones can cross this island. The majority of cyclones moving from coastal to internal areas of Antarctica will not reach the central part of the continent at all. They will be destroyed by a glacial cooling on the way to the continent's centre. Actually, meteorological observations show that warm and wet air masses from margins of the Antarctic continent (mainly, from the Weddell Sea and an Ice Shelf of Ronne) penetrate inland to the South Pole during no more than 10 days per year [125]. Therefore the vast ice sheets of Greenland and Antarctica can suppress effectively short-term local meteorological fluctuations. Glaciers of smaller sizes (the ice cap of New Land; mountain glaciers) cannot influence the synoptic scale waves considerably [255]. Thus they are less suitable for nitrate paleoastrophysical research.

Wind regime is a factor potentially influencing nitrate records in ice appreciably because winds can mix upper layers of snow. However, not everywhere are winds closely linked to local meteorology. Antarctica is a region where strong katabatic winds - air streams caused by the downward motion of cold and dense air, the speed of which is defined by a surface inclination - appear frequently. These winds, weakly connected with weather conditions, are caused by the cooling of air masses over a surface of an ice sheet and their subsequent drain downwards to the coast by gravity. Winds in coastal areas of Antarctica can reach extraordinary forces. For example, at cape Dennison ($\varphi=67.0°$ S, $\lambda=142.4°$ E) on the Commonwealth Seacoast, the yearly-average speed of wind is 19.4 m s^{-1} and its maximum value about 90 m s^{-1} (the strongest winds globally). At a Russian station on a very different part of the continent (74.5° S, 136.5° W), the mean annual wind speed is 13 m s^{-1} and during 136 days per year the speed of wind exceeds 30 m s^{-1}. As a consequence 'zastrugi' - the local peculiarities of surface micro-relief formed by wind and looking like dunes - can reach a height of 1.0-1.5 m. Thus, the characteristic scale of distortions of the local micro-relief caused by wind reaches 1.0 m whereas the thickness of an annual snow layer in coastal parts of Antarctica does not exceed 0.3-0.5 m. In inland areas it does not exceed 0.1 m. Such mixing can average a considerable concentration of isotopes in the top snow layer and effectively smooth their short-term (weeks or months) fluctuations. In central Antarctica, winds are much weaker. At the Vostok station (φ=78.8° S, λ=106.8° E, elevation 3488 a.s.l.), the mean annual speed of wind is 5.4 m s^{-1} and at Amundsen-Scott station (South Pole), about 6.0 m s^{-1}. So, the central and more or less flatter parts of Greenland and Antarctica are the best places for extracting ice samples for nitrate paleoastrophysics due to:

(a)   high-latitude dislocation (low geomagnetic rigidity cutoff, polar night);

(b)   relatively stable meteorological regime;

(c)   relatively low speeds of katabatic winds (upper and flat parts of ice sheet).

The nitrate content in ice samples usually is measured by spectrophotometer. Dating the obtained datasets is based on: (a) apparent seasonal variations in NO_3^- concentration and (b) dated volcanic horizons. The generally higher levels of nitrate concentration [NO_3^-] during summer is a result of increased snow sublimation under sunlight [125] and the facilitation of access to the middle latitudes air masses enriched by anthropogenic and biogenic NO_x to the central areas of Greenland and Antarctica during the warm period. Huge input of sulphides into the atmosphere after each powerful volcanic eruptions results in the electrical conductivity in the ice layer of corresponding year increases sharply. The conductivity, measured along the ice core simultaneously with the NO_3^- abundance, therefore provides researchers with a number of useful time markers. Moreover, the nitrate series often contain the direct fingerprints of major volcano eruptions. Several examples of sharp reductions of NO_3^- concentration coinciding with conductivity and [H_2SO_4] peaks has been reported [125, 251].

3.5. Variations of Nitrate in Polar Ice and their Paleoastrophysical Implementations

3.5.1. Short-Term Variations of NO_3^-

Many researchers [236, 240, 256-258] have reported the existence of an unambiguous connection between SPE and short (less than 2 months) but prominent peaks in nitrate concentration both in Antarctica and Greenland. Evidence for the supernovae of Tycho (1572) and Kepler (1604) were obtained by Dreschhoff and Laird [256]. In addition, 11-year and 22-year cycles in nitrate data were also reported by Zeller and Parker [240] and Dreschhoff and Zeller [236]. Weak but significant increase in mean nitrate concentration after solar proton events was revealed in the Antarctic NO_3^- record by Palmer *et al.* [259]. McCracken *et al.* [260] examined nitrate abundance in a 125.6 m long GISP2 H core retrieved at Summit, Greenland (φ=72°N, λ=38°W, altitude 3210 m) and two shorter cores drilled at Windless Bight (φ=78°S, λ=167°E) on the Ross Ice Shelf in Antarctica. The cores were analysed at ultra-high resolution (1.5 cm intervals). McCracken *et al.* [258] used the dating procedure described above and provided absolute dating of the samples with an accuracy of ±2 months for 90% of the Greenland core, and ±1 year where the annual cycle was poorly manifested. They showed that strong SPE with F_{30} - omni-directional fluence of energetic (E_p>30 MeV) protons - exceeding 10^9 particles per cm^2 can be identified in polar ice. McCracken *et al.* [258] identified 70 strong SPE (F_{30}>2 ×10^9 cm^{-2} between 1561 and 1950). The proton fluences (cm^{-2}) were also assessed. It should be noted that some strong solar proton events in the past were identified using other paleorecords. Usoskin *et al.* [261] found 10 SPE (1755, 1763, 1774, 1793, 1813, 1851, 1867, 1895, and 1927) using the data on concentration of cosmogenic beryllium in Greenland ice. Moreover, Kostantinov *et al.* [262] performed a joint study of the data on ^{14}C and ^{10}Be in similar natural archives. The authors revealed substantial increases of SCR flux during 1750-1790, 1851-1853, 1868-1869, and 1896. These results agree with the results of McCracken *et al.* [258] who identified strong solar proton events with F_{30} > 2×10^9 cm^{-2} in 1755, 1763, 1774, 1793, 1813, 1851, 1866, 1868, 1895, 1896, and 1928. Thus, the study of a number of nitrate series measured with ultra-high (several samples per year) time resolution showed that they contain information about powerful SPE in the past. However, other studies [32, 238, 263] have found no such link. Some authors suggested a minor contribution to the polar ice NO_3^- record from the middle atmosphere in relation to tropospheric sources [32, 238]. Further, SPE fingerprints can be masked by local meteorological or anthropogenic events [263]. Thus the possibility of nitrate record to reflect the short-term SCR fluctuations still is debated.

3.5.2. Mid- and Long-Term Variations of NO_3^-

Evidence for a link between [NO_3^-] and solar variability over longer time scales has been achieved as well. Indeed, Mayewski *et al.* [239], Kocharov *et al.* [145] and Ogurtsov *et al.* [264] have reported the presence of a century-long cycle in the concentration of NO_3^- ions in Greenland ice. Significant correlations between Wolf number and nitrate concentration in two Central Greenland cores, wavelet filtered in the century-scale cycle band (55-147 years), was found between 1700-1980 by Ogurtsov *et al.* [264]. An analysis of the Central Greenland nitrate record by McCracken *et al.* [258] also demonstrated that SPE follow the century-scale

(Gleissberg) periodicity through 1561-1950 (Fig. **3.17**). One can note that during, the period of instrumental measurement of solar cosmic rays (after the mid-20th century) is a time of rather low frequency of SPE occurrence (see Fig. **3.17**).

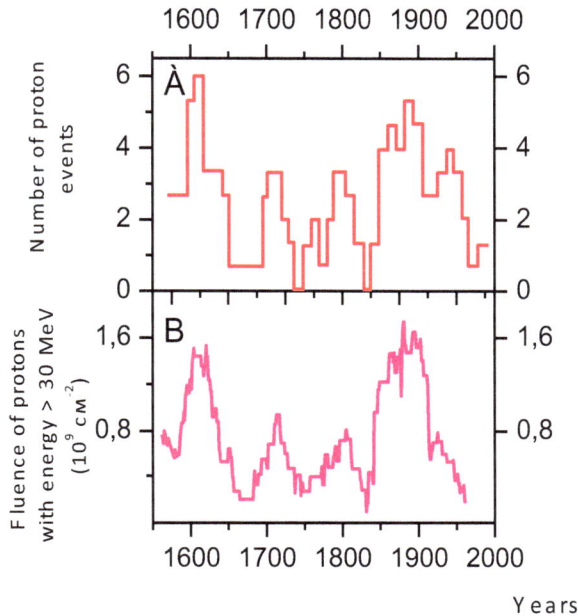

Fig. (3.17). (**A**) the frequency of SPE averaged over two solar cycles (scanned from McCracken *et al*. [258] and digitized); (**B**) the fluence of strong SPE reconstructed by McCracken *et al*. [260] averaged over 30 yrs.

In a recent paper by Traversi *et al*. [130], significant wavelet coherency between reconstructed cosmic ray intensity and nitrate in the Gleissberg cycle band (50-100 years) was revealed in the TALDICE ice core (East Antarctica) during 1713-1913. Since the meteorological conditions of East Antarctica are very different from those in Central Greenland, the results of [130] corroborates that the century-type periodicity is a robust feature of nitrate variability related to the solar Gleissberg cycle, and not to climatic changes. Furthermore Wang *et al*. [265] have found 88-year variation of NO_3^- concentration in a glacier at Tibetan plateau.

Traversi *et al*. [130] also showed that nitrate series covering the Holocene period (645-11 400 BP) have a significant agreement with the radiocarbon-based cosmic ray flux reconstruction on the millennial and multi-millennial time scales. That testifies that ice core nitrate is a perspective proxy of solar activity on the centennial-millennial time scale.

3.6. Historical Naked-Eye Sunspot Observations

The very large (a total area more than 1900 MSH) sunspot groups and sunspots can be observed with bared eye at sunset and sunrise or through smoke and fog [266]. According to Eddy *et al.* [66], such sunspots can appear when mean daily Wolf number exceeds 50. These sunspots were detected during the pre-instrumental epoch by ancient Oriental (particularly Chinese) astronomers. At present the data on ancient sunspot observations made by naked eye (SONE) is the longest direct record of solar activity in the past. The modern catalogues of SONE cover a time interval falling between 165 BC to AD 1918, including more than 200 events [25, 267]. The accuracy and reliability of information on the Sun's activity obtained from historical chronicles has been studied in a number of works (see [50, 266, 268]), showing that Oriental historical data really reflect such important features of solar activity as 11-year, century-long and bi-centennial periodicities, profound Maunder-type minima of solar activity, as well as the butterfly diagrams [26]. Comparison of direct (SONE) and proxy (cosmogenic isotopes, nitrate) data is important for the purpose of extracting increasingly reliable information about the long-term changes of solar activity. The qualifications of ancient Chinese astronomers were also high enough because they made observations of many solar and lunar eclipses, supernovae, novae, comets, meteor showers and even a possible observation of Jupiter's satellite Ganymede (364 BC) confirm their professionalism. On the other hand, SONE record has some considerable shortcomings:

(a)   It is probably that ancient astronomers often mixed actual sunspots with other meteorological and celestial phenomena.

(b)   Ancient solar observations were not systematic and thus they are non-uniform in time. Very often sunspots were reported near the day of new moon. It happened because just in these periods surveillance was more intensive, because the determination of new moon date had a significant calendar purposes [25]. In addition, the dating of naked eye observations is not always quite precise.

(c)   Some meteorological effects may exist in different SONE records [268].

Wittmann and Xu [25] compared more than 20 historical sunspot sightings in Asia with the of European telescopic data from 1848-1918: only one third of the unaided-eye observations are confirmed by Western instrumental records. Willis *et al.* [268] found that two thirds of naked eye observations are corroborated by

Western sources in 1874-1918. In any case, the SONE data can serve a qualitative indicator of sunspot variability over a long time scale and thus be used in a number of works (see [26]).

3.7. Achievements of Paleoastrophysics

3.7.1. Epochs of Prolonged Solar Maxima and Minima

Several reconstructions of solar activity with an annual-resolution of different types of parameters for the last 1000 years have been produced (Fig. **3.18**).

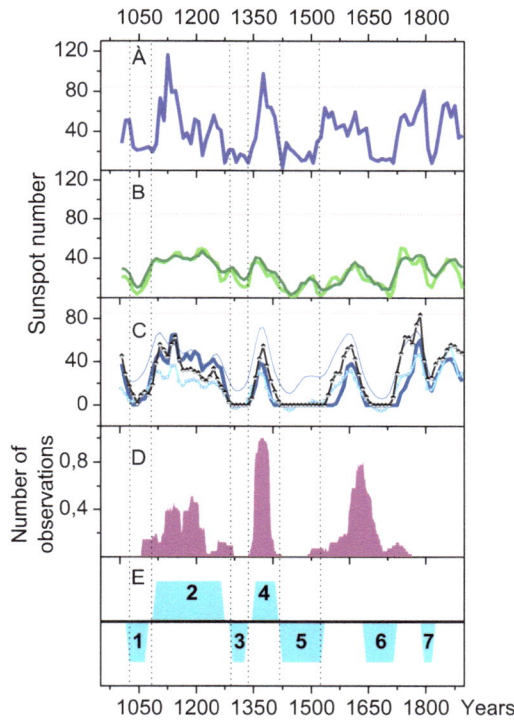

Fig. (3.18). Decadal reconstructions of sunspot number: (**A**) reconstruction after Nagovitsyn based on auroral data [269]; (**B**) reconstructions after Ogurtsov [270] (dark green line) and Usoskin *et al.* [181] (green line) based on ^{10}Be data; (**C**) reconstructions after Nagovitsyn *et al.* [197] (thin blue line), Ogurtsov [196] (thick dark blue line), Solanki *et al.* [176] (cyan line with circles), Stuiver and Quay [144] (black line) based on ^{14}C data. (**D**) number of sunspot observations made by the unaided eye (averaged over 35 years) after Wittmann and Xu [25]; (**E**) 1 Oort (Norman) minimum, 2 Medieval maximum, 3 Wolf minimum, 4 Late Medieval maximum, 5 Spörer minimum, 6 Maunder minimum, 7 Dalton minimum. Red dotted lines depicts the average sunspot number ($R_Z = 85$) over the second part of 20th century.

A comparison of the sunspot reconstructions shows that they agree in key features. The coefficient of correlation between annualy interpolated different proxy-based series varies from 0.52 to 0.92. The correlation between instrumentally measured and reconstructed sunspot numbers reaches 0.70-0.80 over the decadal time scale. Prolonged timespans of very high and very low SA (extreme periods) manifest themselves in all the solar paleoproxies They are demonstrated clearly also by Oort minimum (11th century), Wolf minimum (end of the 13th-beginning of the 14th century), Spörer minimum (the 15th-beginning of the 16th century) and Maunder (1645-1715) minimum as well as medieval (the 12-13th centuries) and late medieval maxima (end of the 14th century) (Fig. **3.18**). Solar maxima are also corroborated by unaided visual sunspot observations. Thus, the existence of extremes of solar activity listed above is confirmed by:

(a) ^{14}C-based sunspot proxies;

(b) ^{10}Be-based sunspot proxies;

(c) sunspot proxy based on auroral observations;

(d) sunspot observations made by naked eye.

Information about the deep minima and maxima of solar activity, their durations, and the time interval between subsequent events has important implications for the development of the solar dynamo theory. For this reason, establishing the deep extremes of solar activity during the last 1000 years may be one of the key achievements of solar paleoastrophysics. Moreover, Usoskin *et al.* [271] found evidence of 27 grand minima and 19 grand maxima of solar activity during the last 11 000 years. Solar paleoproxies show that the contemporary level of solar activity (the last 6-7 decades) is very high - it is likely the highest during the last millennium (see Fig. **3.18**). Solanki *et al.* [176] claimed that it is the highest over the last 8 000 years, although this hypothesis has been disputed by Muscheler *et al.* [272] and Ogurtsov [273]. Thus we are currently living in a time period of grand solar maximum. During grand solar minima, sunspots most likely were absent and the solar modulation parameter Φ dropped below 100 MV (see Beer *et al.* [274]).

3.7.2. Mid- and Long-Term Cycles of Solar Activity

Research of the Sun's variability on a secular time scale is yet another important topic of paleoastrophysics. Since telescopic observations of sunspots started in

AD 1610, direct solar indices have been too short for searching centennial and longer-term solar variability. Despite that, the century-scale cycle of Gleissberg was revealed by means of analysis using the Wolf number though it is not easy to establish this variation absolutely reliably using only sunspot data. An examination of long-scale proxies of the Sun's activity made it possible to unambiguously prove the existence of the century-scale (55-135 years) periodicity of solar activity. Recently, it has been shown that this cycle consists of two periodicities – 55-80 years variation and 90-135 years variation [52]. Bi-centennial (170-260 years) solar variation – the cycle proposed by Suess and de Vries – was also found by analysing proxy data. Other studies on solar reconstructions have also provided plausible evidence for the existence of 500-900-years and ca. 1500-years variations of SA [225, 235, 269]. In addition, ca 2300 years solar cycle has been reported in a few works [275, 276]. This periodicity was called the cycle of Hallstatt because of its correlation with climate epochs documented by Schmidt and Gruhle [277]. The last two variations may presently be considered more or less uncertain (Table **3.5**).

Table 3.5. Solar periodicities established by paleoastrophysical approach.

Solar Periodicity	The Name	Period (Years)	
century-scale	cycle of Gleissberg	55-135	55-85
			100-135
bi-centennial	cycle of Suess	170-250	
quasi 1500 year (?)	cycle of Bond	~1500	
quasi 2000-year (?)	cycle of Hallstatt	1800-2500	

3.7.3. Long-Term Changes of Fluxes of Energetic Solar Particles

Lunar and meteoric rock is another source of information on solar energetic particles in the past (see *e.g.* Usoskin [278]). Measurements of isotope activities in the upper layers of a rock can be used to evaluate the flux of solar cosmic rays integrated over different timescales (Table **3.6**).

We can conclude from Table **3.6** that the mean flux of SCR is relatively stable and the range of its long-term fluctuations over the last 10^6 years was of the same scale as during instrumental era (see also Table **4** in Usoskin [278]). However, short-term pulses of fluxes of solar energetic particles in the past could be rather strong. For example, Usoskin *et al.* [285] examined the AD 775 cosmic ray event,

Table 3.6. Estimations of omni-directional integral fluxes of solar energetic particles with energies above 10 MeV and 30 MeV.

Time Interval	Reference	Method of Estimation	Flux of Protons cm^{-2} c^{-1}	
			E$_p$>10 MeV	E$_p$>30 MeV
1996-2006	Reedy [279]	Satellite measurement	278	61
1986-1996	Reedy [280]	Satellite measurement	152	31
1965-1975	Reedy [279]	Satellite measurement	89	28
1954-1964	Reedy [279]	^{22}Na, ^{55}Fe, lunar rock	378	136
last 5×10^3 yrs	Hoyt *et al.* [281]	lunar rock, Tl	60	14
last 10^4 yrs	Boekl [282]	lunar rock, ^{14}C	200	72
last 2×10^5 yrs	Fink *et al.* [283]	lunar rock, ^{41}Ca, ^{26}Al, ^{10}Be	198	56
last 10^6 yrs	Kohl *et al.* [284]	lunar rock ^{26}Al	70	25

which produced 15 per mille enhancement of ^{14}C content measured in Japanese cedars ($Q =19$ at×cm^{-2}× s^{-1} over a year) using ^{10}Be and ^{14}C series. They found that the event can be explained by an extreme SPE with $F_{30} = 4.5\times10^{10}$ cm^{-2}.

3.8. Remaining Problems

Information of solar activity contained in modern paleoreconstructions based on cosmogenic isotopes is not exact because of the more or less noisy nature of the proxy data. Particularly secular (centennial-to-millennial scale) variations are uncertain. Testing of the quality of modern paleoreconstructions confirms this. A technique for verifying the currently available reconstructions of solar activity in the pre-instrumental era (prior to AD 1615), covering time intervals up to 10 000 years, has been developed by Ogurtsov [273]. The technique is based on investigation of the possibility of predicting the actual sunspot numbers determined by means of telescopic observations using reconstructed solar series. In this approach, the series of group sunspot numbers measured instrumentally in AD 1615-1995 [37] was forecasted using various reconstructed solar series covering periods prior to 1615 as a source of information. A nonlinear forecasting method was used for the prognosis. This method is based on reconstructing the trajectory of the dynamical system of the predicted series in pseudophase space. Testing three of the longest paleoseries (see Fig. **3.10B-D**) showed that they most likely contain only qualitative information about the behaviour of solar activity in the past. In other words, using the currently existing solar paleoreconstructions, it is quite possible for example to reveal periods of global solar maxima, but it is very difficult to assess precisely enough the level of solar activity during each maximum. Accordingly, the hypothesis that the Sun's activity in the 20th century was highest in the last 8000 years [176] may be considered as an assumption that currently can be neither confirmed nor refuted. Further work using new paleoreconstructions is required to clarify this question as well as to increase precision and reliability of our knowledge about past long-term solar variability. Therefore, research involving novel as well as old indicators or proxies ('independent witnesses') of solar activity in the past is represented the basic direction of the further work. For example, in the works of Bonino *et al.* [286] and Usoskin *et al.* [278], the concentration of [44]Ti in meteorites fallen to the ground was considered as a perspective source of data on cosmic ray flux in the past. In addition, a further improvement of models and better descriptions of the redistribution processes of cosmogenic isotopes in terrestrial archives will certainly also be of huge importance.

Concerning nitrate paleoastrophysics, the recent work of Wolff *et al.* [263] assesses and presents the main future challenges in this discipline. The authors examined 13 well-resolved ice core records from Greenland and Antarctica and found no evidence for a sharp spike dating from 1859 (Carrington white flare), which is clearly manifested in a nitrate record of Dreschhoff and Zeller [125, 236]. This result brings up again a question about the character of a link between NO_3^- concentration in ice and astrophysical processes. For example, Wolff *et al.* [263] have claimed that the nitrate record in polar ice cannot serve as an index of solar proton events. It should be noted, however, that a lot of SPE, identified by GISP2 H core nitrate series were corroborated by [10]Be data [261] and ^{14}C data [262] (see Section 3.5.1). Thus, additional and thorough analyses of nitrate variations over different time scales, performed using new experimental series are needed for a definite conclusion. Increasing our knowledge is much in need about processes of nitrate transport as well as changes in deposition and post-deposition.

CHAPTER 4

Palaeoclimatology: Advances and Limitations

Abstract: A general introduction, history as well as major recent advances and problems of paleoclimatology are presented, including paleoclimatology of stable isotopes (^{13}C, ^{18}O), dendroclimatology, and the use of other climate proxies. An estimation of advantages and shortcomings of various sources of the climatic information is made. Moreover, the divergence issue between different proxies is discussed. Long-term climatic oscillations, established by means of paleoclimatology are also described.

Keywords: Dendroclimatology, paleoclimatology, stable isotopes.

Climate is the average pattern of weather for a specified region. *Climatology* is a branch of Earth sciences which focuses on the research of climate regimes of our planet and analyses the causes, character and practical consequences of climatic variations. Till the end of 18th century it was closely connected with meteorology. The concept of climate emerged first in ancient Greece. *Meteorologica* was the first scientific book devoted to atmospheric phenomena - was written by Aristotle (384-322 BC). Hippocrate's (460-377 BC) point of view on climate remained very influential until the beginning of 19th century [287]. The medieval Chinese scientist Shen Kuo (1031-1095) was likely the first to suppose that climate can vary in the course of time. The Enlightenment in Europe opened a new era in both weather and climate research. The invention of meteorological instruments - thermometer (Galileo, 1603), mercury barometer (Torricelli, 1643), barometer-aneroid (Leibnitz, 1700) - opened a new era in meteorology. The nineteenth century saw further advancements in climatology. A. Humboldt (1769-1859) was one of the pioneers of scientific climatology. In 1817 he constructed the first map of global annual isotherms using data from 57 weather stations. Maps of the isotherms of January and July were constructed by H.W. Dove (1803-1879) in 1848. The term anticyclone was invented by F. Galton (1822-1911). Moreover in 1896 S. Arrhenius (1859-1927) famously claimed that the use of fossil fuels would ultimately result in global warming. These and many other discoveries laid the foundation for modern climatology.

Palaeoclimatology is the study of past climates throughout the Earth's history. It reconstructs and studies variations of climate prior to the epoch of instrumental measurements (the pre-instrumental period) by means of various proxy indicators. Proxies are of two basic kinds - natural (physical or biological) records and

Maxim Ogurtsov, Risto Jalkanen, Markus Lindholm and Svetlana Veretenenko

documentary archives. Among the key data used by palaeoclimatology are tree rings (width and density), abundance of stable isotopes ($^{13}C, ^{18}O$, D) in terrestrial archives (ice, tree rings and corals) and historical documents.

4.1. Dendroclimatology

In 1892 P.N. Shvedov compared data on tree-ring growth with data on precipitation from a few meteorological stations in South-East Russia. Shvedov [288] arrived at the conclusion that droughts in the analysed region had an approximate nine-year cycle and highlighted the importance of tree-ring data for further climatic study [289]. A. E. Douglass, who reported the connection between radial tree growth and climatic factors [290], is often considered as the founder of dendroclimatology - the discipline which evaluates past climate conditions from annually-resolved data based on tree growth (usually dated tree rings). The fundamental idea of dendroclimatology is that annual increments of trees, which grow under severe conditions, react detectably to the limiting environmental factors. According to a simplified model by Vaganov-Shashkin [291-293] daily growth rate $Gr(t)$ is a product of the growth rate owed to solar radiation and the minimum of the growth rate owed to either surface air temperature or soil moisture:

$$Gr(t) = Gr_E(t) \times \min[Gr_T(t), Gr_W(t)] \text{,} \qquad (4.1)$$

where $Gr_E(t)$, $Gr_T(t)$ and $Gr_W(t)$ are the daily growth rates resulting from sunlight, soil water balance, and near-surface air temperature, respectively. The partial growth rate $Gr_E(t)$ can be described as:

$$Gr_E(t) = \cos(h_s)\sin(\Phi)\sin(\theta) + \cos(\Phi)\cos(\theta)\cos(h_s) \text{,} \qquad (4.2)$$

where Φ is a latitude, θ is a declination angle and h_s is hour angle [289, 290]. Experimental data (see Evans *et al.* [291] and references therein) show that the growth rates $Gr_T(t)$ and $Gr_W(t)$ can be approximated by trapezoidal waveform functions which consist of three segments: (a) rising growth rates below an optimal range, (b) fairly constant rates within an optimum range, and (c) falling growth rates above the optimal range. In their model representations (Fig. **4.1**) Evans *et al.* [291] estimated temperature critical points as follows: $T_{min}=5°$ C, $T_{opt1}=18°$ C, $T_{opt2}=24°$ C, $T_{max}=31°$ C.

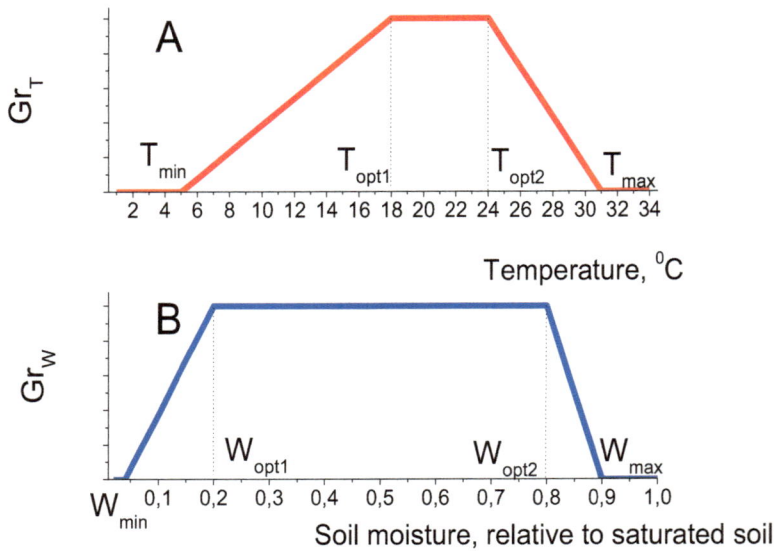

Fig. (4.1). Schematic representation of growth rate functions according to Evans *et al.* [291]. (**A**) growth rate owed to temperature, (**B**) growth rate owed to soil water balance.

Thus, trees growing in extremely cold conditions (northern tree line, upper tree line in mountains) reflect and record variations of growing season temperatures. Dendroclimatical indices - tree-ring width and density of latewood (dense and dark wood, formed in the second half of the vegetative season) - appear to be connected with warm period (summer-time) temperature. Latewood density as a rule reflects temperature change better than tree ring width. Correspondingly *e.g.* the rings of trees living in severely dry conditions may be used as precipitation indicators. Tree rings are one of the most important climate indexes because they can be absolutely dated annually by means of methods of dendrochronology [294]. The high time resolution and accuracy of calendar dating (1 year) is the key advantage of dendrochronology. After Fritts [294], Cook and Kairiukstis [295] accumulated contemporary knowledge and dendroclimatology became popular. Throughout the last few decades methods of dendroclimatology have become major tools in the investigation of past climatic conditions in many parts of the globe (see *e.g.* [296, 300, 301]). Not only radial growth (ring width) reflects changes of climate. Jalkanen and Tuovinen [302] and Pensa *et al.* [303] showed that in Northern Fennoscandia annual height growth as well as needle production of trees were closely related to the mean surface temperature of the previous summer. Thus tree-height and needle-trace records offer a new tool for reconstructing summer temperatures in the past.

Data from tree growth-based indices are usually calibrated against instrumental temperatures in order to make inferences about the past. This allows the relationship between parameters of tree growth and climate to be determined over a calibration period (using *e.g.* a temperature record measured by thermometer at the nearest meteorological station). If a meaningful and time-stable relationship is found (often by linear regression), past temperatures can be estimated with the proxy record. Usually this relationship is tested separately for high and low-frequency components of variability [304], and often even fragmentary early temperature records included for additional verification [305].

Dendroclimatological studies from Northern Fennoscandia (68° to 70° N, 20° to 30° E) have provided reconstructions of past summer temperatures, for example July temperature anomalies (Fig. **4.2**) using samples from living and dead trees:

(a)   Reconstruction by Lindholm and Eronen [306] (negative exponentials and regression lines applied in standardization);

(b)   Reconstruction by Helama *et al.* [307, 308] using tree-ring series (RCS applied in standardization) as well as pollen-stratigraphic data from 11 small lakes situated between 66.25° to 70.50° N and 14.03° to 35.19° E;

(c)   Temperatures measured at Inari meteorological station (69°04' N, 27°07' E) between 1906 and 1991. The coefficient of annual correlation between direct and proxy July temperature records for the period is $r_l = 0.66$ for Lindholm and Eronen's [306] series and $r_l = 0.58$ for Helama's *et al.* [308] series. Standard deviation is ca. 1.5°C for both proxies. The relationships between instrumental temperature from Inari and proxy reconstructions by Lindholm and Eronen [306] and Helama *et al.* [308] are illustrated in Fig. (**4.2A, B**).

Annual (January-December) and growing-season (April-September) temperatures for the extratropical (30° to 70°N latitude) part of the Northern Hemisphere have a correlation of $r_l = 0.87$ for the period 1856 to 2000 on an inter-annual basis and $r_l = 0.93$ after decadal smoothing [298]. The correlations of annual and growing-season 30° to 70°N temperatures from 30° to 70°N with those of the entire Northern Hemisphere (0° to 90°N) are also high. Thus, tree-growth-based chronologies can be used for the estimation of annual hemispheric temperatures. Another advantage of dendrochronological data is that they are usually collected from regions far away from human activities and are less subject to local anthropogenic effects such as urbanization, industry, and changes in landscape.

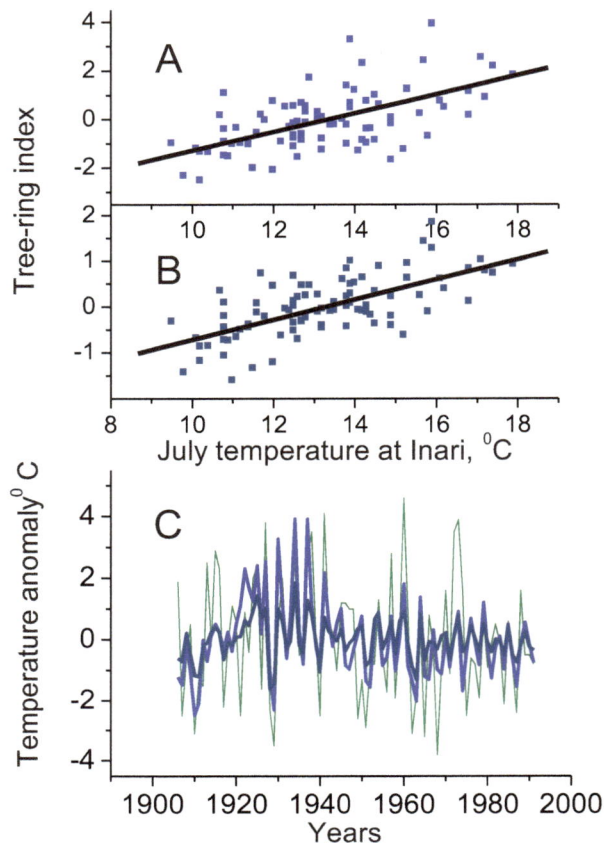

Fig. (4.2). (**A**) Dependence of tree-ring index of Lindholm and Eronen [306] on instrumental temperature from Inari station (69° N, 27° E); (**B**) dependence of tree-ring-pollen index of Helama *et al.* [308] on instrumental temperature; (**C**) anomaly of July temperature over Northern Fennoscandia. Dark blue line - tree-ring based reconstruction of Lindholm and Eronen [306]; violet line - tree-ring and pollen based reconstruction of Helama *et al.* [308]; dark green line - instrumental data. Mean July instrumental temperature over 1906-1991 is 13.3° C.

It should be noted that the calibration period is short because of the limited thermometric data - usually no more than the last 80 to 100 years. This is not long enough to definitely evaluate quality of reconstructed long-term climate variations (periods from centuries to millennia). Besides the shortness of the calibration period dendroclimatology has two other serious shortcomings:

(1)   Suppression of low-frequency (periods of hundred and more years) components of the signal arising from the trend-removal technique (standardization);

(2)   Anomalous reduction in the sensitivity of tree ring growth to temperature over the past few decades of the 20th century.

The first problem is connected with the fact that tree rings have a natural biological trend - a young tree (*e.g.* aged less than 50 years) has wider rings than an older tree. Also the density of late wood has similar trends [297]. The trends can be strong and specific for each tree. That is why the extraction of climatic signals becomes possible only after standardization - removal of long-term biological and other non-climatic components. Clearly, some low-frequency variations linked to corresponding temperature changes may also be removed and lost as well. This obviously limits the applicability of tree-ring records for estimation of long-term temperature variability. Some recent approaches, *e.g.* regional curve standardization (RCS) [298, 309-311], are supposed to preserve as much low-frequency variability in temperature signals as possible. In the RCS method all the samples to be included in the chronology are aligned by cambial age and a regional growth curve is calculated by averaging their values. Then each growth measurement is indexed using the regional curve value for the appropriate ring age and finally the indices are averaged to a chronology.

An anomalous reduction in the sensitivity of tree growth to temperature variability (ARS) has been identified in many tree-ring and latewood density records during the last decades of the twentieth century (see *e.g.* [296, 312-314]). Obvious underestimation of recent warming in dendroreconstructions is also known as the 'divergence problem' (Fig. **4.3**).

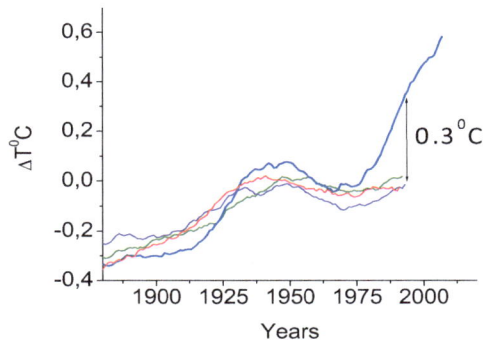

Fig. (4.3). Blue line - temperature of extra-tropical Northern Hemisphere measured by thermometers (http://data.giss.nasa.gov/gistemp/tabledata/ZonAnn.Ts+dSST.txt), violet line - proxy-based reconstruction of Jones *et al.* [315], red line - proxy-based reconstruction of Briffa [296], green line - proxy-based reconstruction of Esper *et al.* [298]. All the data sets are smoothed over 15 years.

According to D'Arrigo *et al.* [313] potential causes of the divergence may be found in both regional and global scale. Among the most probable regional factors responsible for the ARS are:

(a)   Drought stress caused by temperature potentially occurring through the last decades over some boreal and sub-boreal regions, *e.g.* over Alaska [313, 316].

(b)   Delayed melting of snowpack leading to later start of growing season [317] over North Siberia.

The large-scale factors probably responsible for ARS include:

(a)   An increase in the flux of ultraviolet radiation reaching the surface due to decrease in stratospheric ozone concentration which leads to a corresponding decline in tree productivity.

(b)   Global "dimming" - a decrease of surface solar irradiance between 1961 and 1990 [318, 319]. This decrease in solar radiation reaching the ground might result in suppression of photosynthesis and a reduction in tree-ring growth.

(c)   Nonlinear growth reaction, which causes a decrease in tree-ring width under high temperatures [320] (see Fig. **4.1A**).

Despite feasible explanations the ARS problem has still not been solved. Thus, modern temperature dendroreconstructions usually may not cover the last two decades with confidence.

Despite some limitations tree growth-based proxies are an irreplaceable tool in studies of past temperature variability. During the calibration period the coefficient of correlation between temperature series measured by thermometers and dendroreconstructions of temperature can reach 0.4 to 0.5 over the annual timescale and more than 0.8 over the decadal timescale. For latewood density these values range between 0.5 and 0.9 and for tree-height increment proxy between 0.61 and 0.72. The longest available temperature dendroreconstructions cover up to seven or eight millennia.

4.2. Stable Isotope Palaeoclimatology

Urey [321] has suggested that the isotopic $^{18}O/^{16}O$ ratio of carbonates could be used as a palaeo-thermometer. Later Dansgaard [322] put forward the idea of using the isotopic composition of glacier ice as a temperature indicator. Natural geochemical and geophysical processes may result in isotope fractionation depending on temperature because isotopes with different masses have different rates of chemical reactions. As a result it is reasonable to expect the amount of stable isotopes ^{18}O, ^{13}C, and D in terrestrial archives - polar and mountain ice, tree-ring cellulose, corals, bottom sediments - to store information about past climates. Stable isotope palaeo-records, measured by accelerator mass spectrometry, play an important role in modern climatology.

4.2.1. ^{18}O in Polar Ice

Isotope fractioning of the oxygen follows a chain: ^{16}O is lighter than ^{18}O and thus water vapour from the tropical ocean is poorer in ^{18}O than oceanic water. Moving toward high latitudes water vapours lose the heavier and more easily condensed and precipitated ^{18}O water. Thus, the amount of ^{18}O relative to ^{16}O in the water vapour decreases as it approaches the poles, losing mostly ^{18}O water in the form of rain and snow. Likewise, the degree of isotope fractionation depends on temperature - higher temperatures result in higher concentration of a heavy isotope in polar ice and *vice versa*. Sites for ice core drilling are usually chosen close to an ice divide, where the main strain direction is vertical. The ratio $^{18}O/^{16}O$ is determined by means of gas-source isotope-ratio mass spectroscopy. The results of measurements are expressed as $\delta^{18}O$ which is per mille (0.1%) divergence from the standard mean ocean water (SMOW) ratio $(^{18}O/^{16}O)_{SMOW} = 20052 \times 10^{-3}$:

$$\delta^{18}O = \left[\frac{(^{18}O/^{16}O)_{sample}}{(^{18}O/^{16}O)_{SMOW}} - 1 \right] \cdot 1000 \, ‰, \tag{4.3}$$

The precision of modern measurements reaches 0.2‰ [323]. The isotopic composition of $^{18}O/^{16}O$, measured in polar and mountain ice, is one of the most widely used palaeo-indicators of temperature. The relationship between contemporary yearly-averaged temperatures at surface (T °C) and $\delta^{18}O$ in precipitation can be roughly approximated using a simple linear dependence:

$$\delta^{18}O = (\alpha T - \beta) \, ‰, \tag{4.4}$$

where both coefficients α and β are spatially variable: over Greenland $\alpha = 0.6$-$0.7‰/°C$ and over Antarctic $\alpha = 0.44$-$0.84‰/°C$ [322, 324, 325]. Dansgaard [322] estimated β as13.6‰. At the Delingha site (Tibetan plateau, 37°22' N, 97°58' E, 2981 m a.s.l.) $\alpha = 0.638$ and $\beta = 13.31$ [323].

Natural stable isotope time series have a seasonal variation which is useful for annual dating. Going deeper into the ice the yearly layers become thinner and in the deepest part they are indistinguishable. Consequently, the dating of the deeper layers depends appreciably on the model flow processes. Timescale uncertainty is the most important source of error in ice-core-based reconstructions. The identification of annual layers is more difficult at sites where snow accumulation rates are less than 100 mm of water equivalent per year. The major part of ice-core palaeo-isotope series is obtained at locations situated far from any meteorological stations. Therefore many of these time series are not calibrated against instrumental temperature. Another essential limitation is connected with the dependence of ^{18}O concentration in polar ice not only on local temperature during a snowfall but also on temperature at the location of vapour generation and on the distance between a moisture source and a place of snowfall. That is why long-term changes in sea-ice cover (distance to vapour source location) and storm track position could influence the $\delta^{18}O/T$ gradient in polar ice as well.

Records of stable isotope concentration in Greenland ice span time intervals up to 250,000 years and those from Antarctica up to 750,000 years. Correlation between the temperature reconstructed from $\delta^{18}O$ in ice core and instrumental records reaches 0.3 to 0.4 over the annual timescale. The uncertainty of estimation of mean annual temperature using ^{18}O records in certain years could reach $6.0°$ C in Antarctica $(\bar{T}_{ann} \cong -50°C)$ and $2.5°C$ in Greenland $(\bar{T}_{ann} \cong -30°C)$ (see Fig. **4.1** in [325]).

Ocean coral skeletal rings or bands and bowls of molluscs are also known to impart information about climatic conditions in the past. Annually banded corals from the tropical and subtropical regions provide proxies of the past environmental variability of the ocean surface. Lower temperatures as well as denser water salinity cause coral to use heavier isotopes in its structure. But the connection between coral palaeo-records and temperature is usually weaker than in ice-core records as changes of salinity, solar radiation and nutritious ability of the environment play a considerable role. Deep ocean sediments have been used as a source of information about climatic conditions over the past 1 000 000 years.

4.2.2. ^{13}C in Tree Rings

The isotopic composition of wood cellulose is another field allowing for climate reconstruction. Stable isotope $^{13}C/^{12}C$ ratios in organic matter are determined by biological and physical processes during photosynthetic uptake of CO_2 from the air [326]. The abundance of ^{13}C in tree-ring cellulose is expressed as $\delta^{13}C$ values relative to the Vienna Pee Dee Belemnite standard $(^{13}C/^{12}C)_{PDB}$ = 0.01112372. The precision of measurements reaches 0.10-0.13‰ [327, 328]. The fact that the $\delta^{13}C$ value in trees is dependent on climate was revealed over thirty years ago [329, 330]. The mechanism of the ^{13}C-climate relationship is intricate and depends on many factors including very local environmental conditions. In spite of such complications, however, an appreciable connection between $\delta^{13}C$ in tree rings and climatic parameters such as growing season temperature, precipitation regime, relative air humidity, and sunshine hours have been established by many researchers [331-335]. A distinct response of $\delta^{13}C$ in Northern Finland to mid-summer temperature was revealed by Hilasvuori *et al.* [327] and McCarroll *et al.* [333]. The coefficient of correlation between the $\delta^{13}C$ record from Kessi (68°56' N, 28°19' E) obtained by Hilasvuori *et al.* [327] and July temperature measured at the weather station at Inari from 1906 to 1995 is 0.68 on the annual timescale and 0.81 on the decadal timescale. A stable carbon record, obtained from Laanila research area (68°28' to 68°30' N, 27°16' to 27°27 E) by McCarroll *et al.* [333], correlates with mean July-August temperature measured at Sodankylä meteorological station (67°22' N, 26°37 E) from 1958 to 2002, with r_1 = 0.72 on the annual scale. Standard deviation between the Laanila proxy and instrumental records is 0.68° C [333]. Thus $\delta^{13}C$ tree-ring records with annual resolution are important for further palaeoclimatic research.

4.3. Borehole-Based Temperature Estimations

Subsurface terrestrial borehole temperature profiles can be used for assessment of ground surface temperature variations back in time. The estimations are carried out using certain simple suppositions about: (a) the physical properties of the earth near the borehole and (b) a link between surface air temperature and ground surface temperature. The measurements of borehole temperatures are the only source of direct information about past temperatures. A main benefit of geothermal data as compared to those from many other climate proxies (tree rings, ice cores, corals, and historical documentary records) is that they do not need calibration against independent surface instrumental data. In contrast with tree rings, borehole temperatures reflect only climate variability on multi-decadal or longer timescales because of attenuation of short-term fluctuations by the heat diffusion process. Geothermal temperature-depth data have lately been used to estimate changes of Northern Hemisphere continental temperature over the past 500 years [336-339] (Fig. **4.4**).

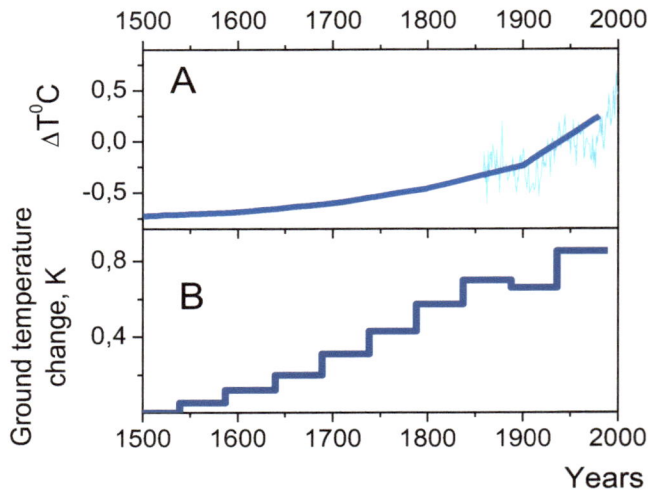

Fig. (4.4). Change in global surface temperature. Cyan line - instrumental data, blue lines - temperature reconstructions from geothermal data. A - Rutherford and Mann [338]. B - Beltrami [333].

In their series Rutherford and Mann [338] analysed 453 individual Northern Hemisphere borehole temperature records, and Beltrami [336] 826 temperature-depth boreholes profiles distributed worldwide (we obtained this data set by scanning and digitizing Beltrami's [336] Fig. **4.4**). These geothermal reconstructions are plotted together with the instrumental temperature data [322]. It should be noted that factors independent on temperature such as changes in land cover, soil moisture and winter snow cover affect the sub-surface thermal properties [339]. Uncertainty of the borehole-based temperature was estimated as 0.24°C during 1500 to 1600 and 0.13°C after AD 1600 [337].

4.4. Historical Data

The observations of different meteorological and phenological phenomena performed by contemporaries and described in various written sources (annals, diaries, archival records) also represent considerable information on past climate variability. These include extreme weather phenomena (droughts, river floods, severe winters, frosts and snowfalls in the summer) as well as the dates of first frosts, the freezing of rivers and lakes, the harvest yields, and flowering dates. The flood level record of the Nile River is probably the longest direct instrumental climatic series available today. It covers the period from AD 640 to AD 1891 [340, 341]. Other historical data obtained in Europe, China, Japan and Russia cover time intervals up to the last 1000 years. Based on Russian chronicles Borisenkov and Pasetskiy [342] calculated a smoothed index of Russian winter severity for AD 1000 to 1900 (Fig. **4.5**). The value of the index in a year *t* was considered as +1 if there was chronicled evidence that the winter was mild, - 1 if there was evidence that the winter was abnormally severe and 0 in all other cases. The index of Russian winter severity has variations of periodicities of ca. 100 years and ca. 200 years.

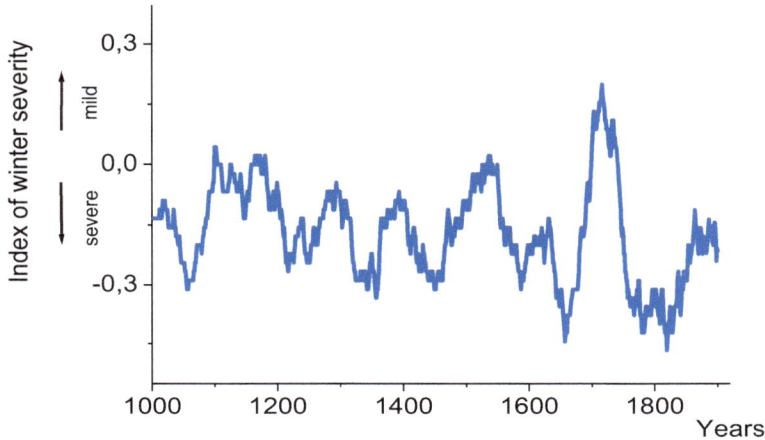

Fig. (4.5). Index of severity of winters in Central Russia (ca. 50 to 60° N, 30 to 40° E) averaged over 45 years according to the chronicle data of Borisenkov and Pasetskiy [342].

It should be noted that only a few of the historical data have been calibrated against instrumental temperature data. Historical data are often highly subjective and inhomogeneous. Thus, extraction of information on changes of climatic parameters over timescales exceeding the duration of a human life - generalizations of the data obtained by several observers - is usually rather difficult [299]. However, perspective methodology has been developed to produce centennial records on past climate extremes [343].

4.5. Ice Core Melt Layers

Ice core melt layers hold information about summer temperatures. For coastal Greenland and glaciers located in places where the summer temperatures rise above freezing at least for some days it is possible to estimate summer temperature from the thickness of the melted layer. According to Krenke [344], summer snow melting can be assessed with the formula:

$$A = (T + 9.5)^3 , \tag{4.5}$$

where A is width of melt snow in mm of water and T is mean June-August temperature in degrees Celsius. Tarussov [345] reported that at Svalbard the thickness of summer melting is 15 to 25 cm (yearly precipitation 50 to 70 cm) and at Novaya Zemlya 60 to 80 cm (annual precipitation 80 to 100 cm). Refrozen melt layers, which are dense and have no air bubbles, are very different from the other layers and thus can be easily distinguished. It should be borne in mind that the data on melt crust width might not capture the full range of the temperature variability - some very cold summers could cause no thaw whereas very warm summers could result in thawing of the whole layer. For this reason the coefficient of annual correlation between melt layer proxy records and temperature, measured by thermometers, is sometimes less than 0.2.

Besides the proxies described above, palaeoclimatology uses other natural archives as indicators of climatic change in the past, *e.g.* gas content in ice core air bubbles, pollen, stable isotopes in cave stalagmites, marine and lake varves, *etc.* The record of atmospheric gases preserved in the form of air bubbles, captured within the polar ice, could also carry important climatic information. Surface snow has a density of ca. 0.35 $g \times cm^{-3}$. The density of firn - granular, partially consolidated snow, which is still full of pores - is 0.55-0.83 $g \times cm^{-3}$. The maximum density of solid ice is about 0.92 $g \times cm^{-3}$. The transformation of firn to ice occurs over density ranging from 0.80 to 0.83 $g \times cm^{-3}$. After density of firn reaches 0.83 $g \times cm^{-3}$ all the air bubbles become closed and any connection between the trapped air and the surface ceases. The profile of snow density in Antarctica can be approximated by a linear equation [346]:

$$\rho = 7.46 \times 10^{-5} H + 0.35, \tag{4.6}$$

where ρ is in $g \times cm^{-3}$ and depth H in cm. Depth of firn-ice transition is: 65 to 70 m at Dye-3 (South Greenland), 95 m at Vostok station, 115 m at South Pole, 77 m at GISP2 site (Central Greenland), and 13 m at glacier of Mount St Elias [347].

Pollen grains, which are extremely resistant to decay, are often washed or blown into lakes and peatlands and accumulate in sediments. It is also possible to recognize a plant species from its pollen grain. The identified plant population of the region at the corresponding time from that sediment layer can inform researcher about the prevailing climatic condition. Pollen-based records are considered sensitive to variability over multi-decadal and longer time scales.

Pollen analysis has been used to extract quantitative information about the climate of the past 200 000 years.

4.6. Multi-Proxies

The degree to which the different palaeo-proxies and instrumental temperatures relate varies (Table **4.1**). The time period during which the correlation was

Table 4.1. **Comparison between proxy and instrumental series. Proxy data are taken mainly from Jones *et al.* [299] and Barnett *et al.* [348]). Correlation is calculated on both annual and decadal scales between the proxy data and instrumental temperature related either to the corresponding grid-box or closest single station (marked with an asterisk).** MXD = maximum latewood density, TRW = tree-ring width, INS = instrumental measurements, HIS = estimation based on historical documents, MI = melt ice index.

Location	Coordinates	Palaeo-Type	Source	R_l (1 yr)	R_l (10 yrs)	Time Interval (Years)
N. Fennoscandia	68° N., 22° E	MXD	Briffa *et al.* [309]	0.79	0.80	100
N. Urals	66° N, 65° E	MXD	Briffa *et al.* [310]	0.81	0.92	100
Yasper	52° N, 117° W	MXD	Luckman *et al.* [350]	0.48	0.45	96
W. USA	45° N, 115° W	MXD	Briffa *et al.* [351]	0.60	0.79	100
N. Fennoscandia	68°- 70° N, 20°- 30° E	TRW	Lindholm and Eronen [306]	0.66*	0.79*	85
N. tree line	58° N, 95° W	TRW	D'Arrigo and Jacoby [352]	0.34	0.87	75
Tasmania	42° S, 146° E	TRW	Cook *et al.* [311]	0.42	0.58	10
Lenka	41° S, 72° W	TRW	Lara and Villalba [353]	0.36	0.55	82
Central England	52° N, 2° W	INS	Parker *et al.* [354]	0.84	0.80	100
Central Europe	46° N, 8° E	HIS	Pfister [355]	0.90	0.83	99
N. Fennoscandia	68°56' N, 28°19' E	$\delta^{13}C$ (tree ring)	Hilasvuori *et al.* [327]	0.68*	0.81*	90
Crete (Greenland)	71° N, 36° W	$\delta^{18}O$ (ice)	Fisher *et al.* [356]	0.30	0.49	99
Law Dome (Antarctica)	67° S, 112° E	$\delta^{18}O$ (ice)	Morgan and van Ommen [324]	0.26	0.98	25
Svalbard	79°N, 15° E	MI	Tarussov [345]	0.08	0.38	82
Galapagos	0.0°, 91° W	$\delta^{18}O$ (corals)	Dunbar *et al.* [357]	0.39	0.16	92
New Caledonia	22° S, 166° E	$\delta^{18}O$ (corals)	Quinn *et al.* [358]	0.41	0.48	85

calculated varies (Table **4.1**). Generally climatic proxies account for 20 to 60% of the annual variance and thus apparently underestimate the actual levels of natural variability [348]. The main features of the different kinds of temperature palaeo-proxies, including approximate estimations of the maximum coefficient of correlation between proxy and instrumental records vary markedly (Table **4.2**). A detailed description of the palaeo-indicators can be found in [296, 299, 348, 349].

Table 4.2. Characterization of different palaeo-proxies of temperature.

Proxy Variable	Maximum Time Span	Spatial Limitations	Maximum R₁ (1 yr)	Maximum R₁ (10 yrs)	General Shortcomings
Tree-ring density	1.5-2 millen-nia	Extratropical (> 30°) part of the Earth or high elevation area	0.79	0.92	Standardization methods complicate interpretation of multi-decadal and longer variability. Divergence problem.
Tree-ring width	7-8 millen-nia	Extratropical (> 30°) part of the Earth or high-elevation area	0.66	0.80	Standardization methods complicate interpretation of multi-decadal and longer variability. Divergence problem.
Tree-height increment	1263 years	Northern Fennoscandia	0.61	0.72	A novel approach not examined in depth.
Stable isotopes in tree rings	1-2 millen-nia	Extratropical (>30°) part of the globe or high elevation area	0.68	0.81	A complicated procedure of measurement.
Stable isotopes in ice	7.5×10^5 years	High-latitude (>60°) and high-elevation ice caps	0.30	0.50	A complicated procedure of measurement. Problems with calibration regarding instrumental data. Problems with dating deep layers.
Stable isotopes in corals	10^6 years	Tropical (±30°) oceans	0.41	0.66	A complicated procedure of measurement. Problems with interpretation of multi-decadal and longer variability (changes in water depth, nutrient supply).
Contemporary written historic records	1372 years	Europe, China, Japan, Korea, Russia, Egypt	0.90	0.93	Problems with interpretation of variability longer than a human lifetime.
Borehole temperature	2×10^4 years (usually - 500 years)	Mid-latitude (30-60° N) part of Northern Hemisphere, southern (>0° S) part of Africa, extracon-tinental Australia	–	–	Reflect only long-term (century-scale and longer) variability.
Ice core melt layers	600 years	High-latitude (>60°) and high-elevation ice caps where temperature in summer reaches positive values	0.18	0.74	Problems with calibration regarding instrumental data. Could not reproduce the whole range of the temperature variability.

Since different palaeo-indicators reflect more or less different features of actual (measured) temperature changes, *multi-proxies* - data sets which generalize proxy records of different types - are often used in palaeoclimatology. For example, Mann *et al.* [359] obtained their reconstruction using twelve proxy data sets - seven based on tree-ring width, two on latewood density, two on $\delta^{18}O$ in ice core, and one on ice accumulation rate. McCarroll *et al.* [360, 361] used multi-proxy analysis to combine several types of growth-based parameters measured from the same trees at the northern timberline.

Moberg *et al.* [363] combined proxies of high temporal resolution (dendrochrono-logical data) and low temporal resolution (borehole data, stalagmite layer thickness pollen, sediments,) using the wavelet technique. Several up-to-date reconstructions of temperature provide information about climate history through the past 1000 years (Table **4.3**). Each of the reconstructions (Fig. **4.6**) was scaled by Ogurtsov *et al.* [365] so that its mean value, averaged over the time interval 1880 - 1980, was equal to the respective mean instrumental temperature of the extratropical (25° to 90° N) part of the Northern Hemisphere (http://data.giss.nasa.gov/gistemp/tabledata/ZonAnn.Ts+dSST.txt). This adjustment was performed in

Table 4.3. **Millennial-length reconstructions of the Northern Hemisphere temperature. Correlations were calculated with the instrumental extratropical temperature for 1880 to 1980 using both raw and 11-year smoothed values (annual and decadal timescales).**

Source	Abbreviation	Time Span	R_1 (1 Year)	R_1 (10 Years)	Proxies Used
Mann *et al.* [359]	NHM99	1000- 1980	0.82	0.96	Tree-ring, $\delta^{18}O$, melt ice
Jones *et al.* [299]	NHJ	1000- 1991	0.67	0.87	Tree-ring, $\delta^{18}O$, melt ice, historical data
Crowley and Lowery [362]	NHC	1000- 1993	-	0.76	Tree-ring, $\delta^{18}O$, pollen, historical data
Esper *et al.* [298]	NHE	850- 1992	0.69	0.95	Tree-ring
Briffa [296]	NHB	1- 1993	0.54	0.86	Tree-ring
Moberg *et al.* [363]	NHMb	1- 1979	0.34	0.54	Tree-ring, $\delta^{18}O$, pollen, Mg/Ca, diatoms, stalagmite, borehole
Loehle [364]	NHL	16- 1980	-	0.82	$\delta^{18}O$, pollen, di-atoms, Mg/Ca, stalagmite, historical data
Mann *et al.* [301]	NHM08	200- 1995	-	0.95	Tree-ring, variety of natural proxy records of different kinds, historical data

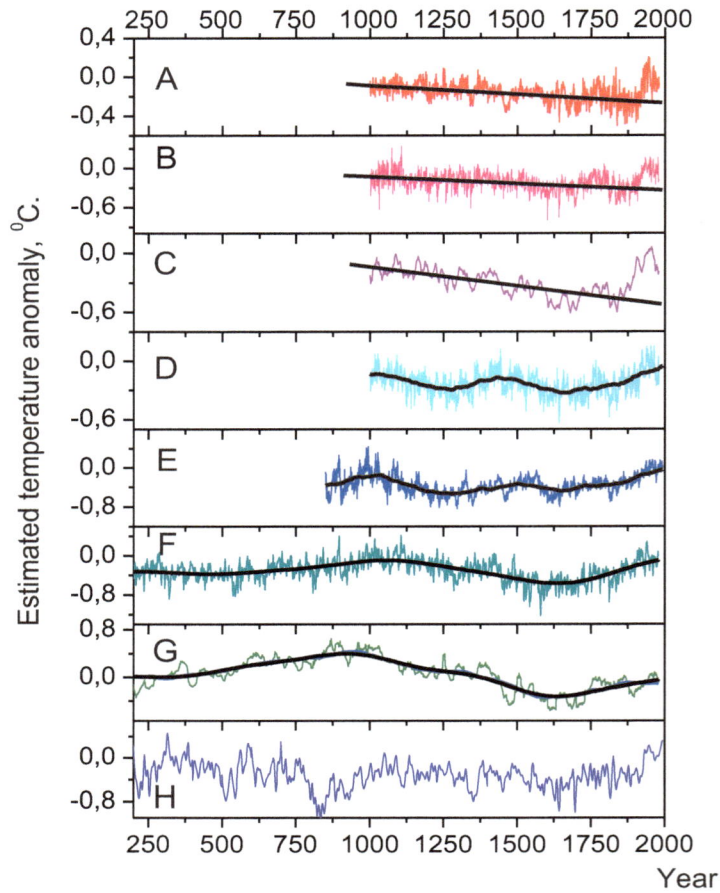

Fig. (4.6). Millennial-scale temperature reconstructions obtained by: (**A**) Mann *et al.* [359]; (**B**) Jones *et al.* [299]; (**C**) Crowley and Lowery [362]; (**D**) Briffa [296]; (**E**) Esper *et al.* [298]; (**F**) Moberg *et al.* [363]; (**G**) Loehle [364]; (**H**) Mann *et al.* [301]. All the time series were adjusted to the mean for 1880-1980 ($\overline{T}$ = -0.125° C). Black lines - long-term trends.

order to facilitate comparison of different reconstructions and took into account that very few temperature indices from tropical regions were used in building the multi-proxies.

The multi-proxy reconstructions (Table **4.3**) account for up to 65% of the annual variance and up to 90% of decadal variance of instrumental records. The standard deviation between multi-proxy and instrumental temperature records (Fig. **4.7**) is 0.15 to 0.25 °C over the annual scale for 1880 to 1980.

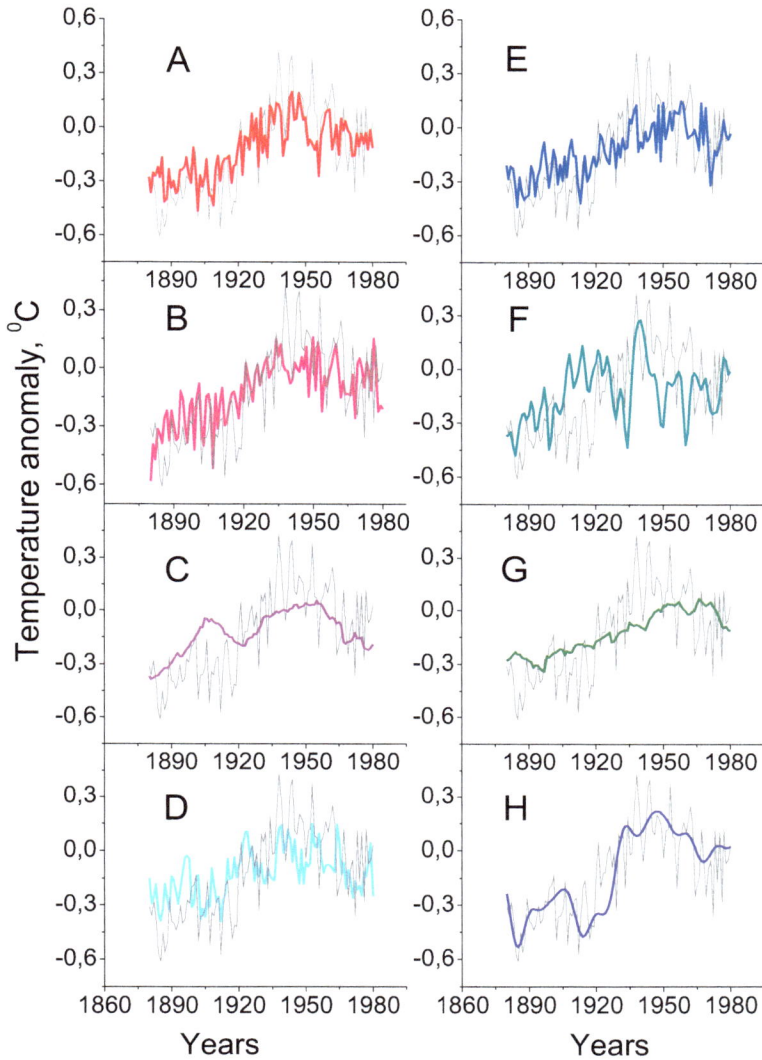

Fig. (4.7). Instrumental temperature of extra-tropical part of the Northern Hemisphere (thin grey line: http://data.giss.nasa.gov/gistemp/tabledata/ZonAnn.Ts+dSST.txt). Coloured lines - temperature reconstructions after: (**A**) Mann *et al*. [359]; (**B**) Jones *et al*. [299]; (**C**) - Crowley and Lowery [362]; (**D**) Briffa [296]; (**E**) Esper *et al*. [298]; (**F**) Moberg *et al*. [363]; (**G**) Loehle [364]; (**H**) Mann *et al*. [301].

4.7. Achievements of Palaeoclimatology

Analysis of a variety of proxies of past climates made it possible to establish reliably climatic oscillations for periods from 10^3 to 10^5 years. These variations, particularly for the ca. 100-kyr glacial-interglacial cycle, have the largest amplitudes and hence result in the most prominent natural changes. Increasing knowledge about climatic variability over the Quaternary period obtained by palaeoclimatology is one of the important achievements of this science over the last few decades.

4.7.1. Astronomical Climatic Cycles

The climate documentation during the last 430 kyr is the result of long-term scientific collaboration between Russia, the USA and France at the Russian Vostok station in Central Antarctica (78.8° S, 106.8° E, 3488 m a.s.l., mean temperature -55.8°C). A study of the chronology of Antarctic ice cores using different temperature proxies testifies that during this time interval major climatic variations were influenced by insolation as proposed by the Milankovitch hypothesis. Serbian mathematician Milankovitch [366] applied the ideas of J. Croll, suggesting that changes in the intensity of solar radiation received from the Earth were the result of three fundamental variations caused by the gravitational forces of the other planets:

(1)   Change of the Earth's orbit from being almost circular to slightly elliptical (eccentricity). This variation has a period of around 100,000 years.

(2)   Change of the angle of tilt of the Earth's axis from 22.1° to 24.5° (obliquity). This variation has a period of 41 000 years.

(3)   Change of the direction of the tilt of the axis (precession) with periods of 19,000 and 23,000 years [367].

Deuterium and oxygen palaeo-records were obtained from Vostok ice core, drilled down to a depth of 3623 m [368], and the data on summer insolation at 65° N is from Berger and Loutre [369] (Fig. **4.8**). The accuracy of the glaciological timescale is better than 15 kyr for the entire time series and better than 5 kyr for the last 110 kyr [368]. Both δD and $\delta^{18}O_{atm}$ series have appreciable variability in eccentricity, obliquity and precession frequency bands. Similar variations were found in many other palaeo-records [370, 371]. Moreover, the connection between the $\delta^{18}O$ of atmospheric oxygen (Fig. **4.8B**) and 65° N summer insolation (Fig. **4.8C**) is also quite obvious. The relationship between $\delta^{18}O_{atm}$ and mid-June insolation at 65°N was established in [365]. An agreement between astronomic temperature cycles could be considered as validation of the Milankovitch hypothesis, which suggests that ice ages were triggered by minima in summer solar radiation near 65°N, allowing winter snowfall to continue all year and hence to build vast Northern Hemisphere ice sheets. The amplitude of the glacial-interglacial change in the surface temperature is estimated as 10 to 12° C [368]. Knowledge about strong and long-term climatic variability in the past is of high importance for the projection of future climates. For example, an analysis of Vostok and EPICA (European Project for Ice Coring in Antarctica) Dome C ice cores suggest that the new next glacial age unlikely to start within the next 30 kyr [372, 373]. Furthermore, past glacial cold periods provide context and means for clarification of the reaction of the climate system to strong radiative perturbations.

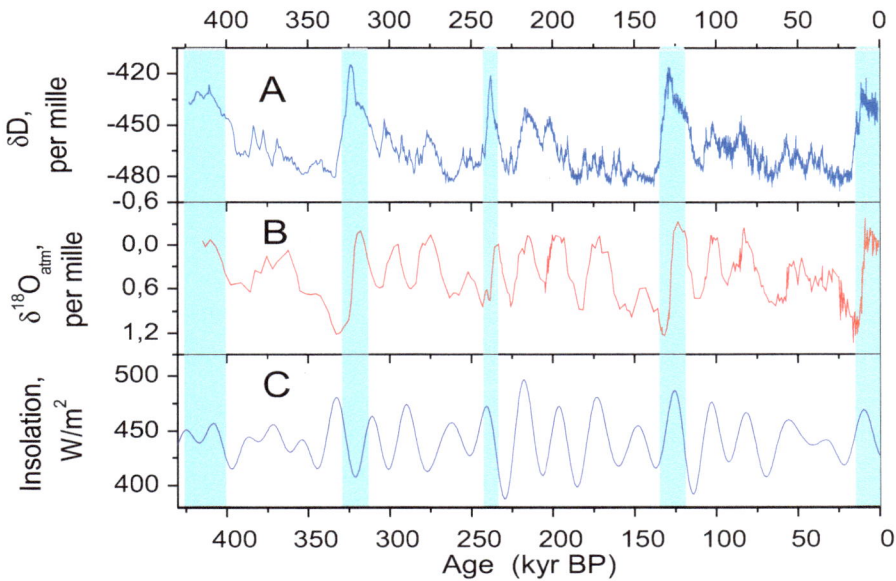

Fig. (4.8). Vostok palaeo-records and insolation: (**A**) the deuterium content of the ice δD; (**B**) $\delta^{18}O_{atm}$ of atmospheric oxygen trapped in bubbles of ice; (**C**) June insolation at 65° N. The shading indicates the interglacial warm periods.

4.7.2. Dansgaard-Oeschger Events

The discovery of Dansgaard-Oeschger (D-O) events - the abrupt climate changes from stadials (cold conditions) to interstadials (warm conditions) - is another milestone of palaeo-isotope thermometry. These events occurred more than 20 times during the last glacial period (roughly 1.5 to 3 ka variation) and were discovered by means of analysis of $\delta^{18}O$ value, measured in the ice core retrieved from Greenland [374]. In the Northern Hemisphere, D-O events take the form of rapid warming episodes with a typical duration of three to four decades, each followed by substantially slower cooling during many centuries. The amplitude of warmer times in Greenland has been assessed as 8 to 15 °C [375]. The sequence of stadial-interstadial transitions is illustrated by the oxygen isotope record from the Greenland Ice Sheet Project 2 (GISP2) ice core (72.6° N, 38.5° W, 3200 a.s.l.) (Fig. **4.9**).

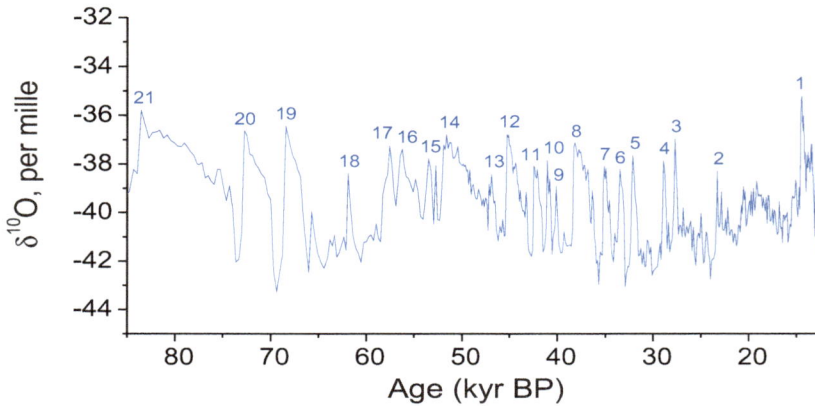

Fig. (4.9). $\delta^{18}O$ record from Greenland (GISP2 ice core [376, 377]). Numbers above $\delta^{18}O$ maxima denote the interstadial (warm) events [378, 379].

Although the effects of the D-O events are expressed more distinctly in ice cores taken from Greenland, there is evidence to suggest that they are globally distributed. An absence of synchrony between D-O events in the two hemispheres [380] indicates that unlike the glacial-interglacial cycles, the D-O events most likely do not involve substantial variations in global mean temperature. The stadial-interstadial changes are probably associated with latitudinal shifts in oceanic convection.

4.7.3. Century-Scale Cycle in Northern Hemisphere Temperature

A global hemispheric-scale spatiotemporal mode of temperature variability on timescales of 50 to 150 years has been found for the Northern Hemisphere by means of study of a network of up-to-date palaeo-reconstructions [381]. On the other hand, Mahasenan *et al.* [382] examined several proxy data sets and found periodicities corresponding roughly to 160 and 80 years. A study of palaeo-data for North America and Europe also showed the existence of century-type variability, which consists of two modes - multidecadal (60 to 80 years) and centennial (>100 year) variations [383]. However the century-long climatic oscillation established in the works above is highly variable geographically, and its spatial pattern is variegated. It is evident from Fig. (**5**) (after Mann *et al.* [381]) that the phases and amplitudes of century-scale variations in individual records diverge appreciably even for neighbouring regions. An analysis of six millennial-length hemispheric-scale temperature proxies by Ogurtsov *et al.* [384] obtained by means of both Fourier and wavelet approaches indicated that besides this regional-scale variability a global century-long climatic rhythm was present in the Northern Hemisphere during the last millennium. The rhythm most likely consists of two periodicities of 50 to 80 and 100-135 years. The century-type cycles in eight temperature reconstructions were selected by means of wavelet filtering using the 55- 147-year band and MHAT (Mexican hat) basis (Fig. **4.10**). The coefficients of correlation between wavelet-filtered data sets were calculated from AD 1000 to 1900 (prior to anthropogenic warming) (Table **4.4**).

There is a clear synchrony between century-long variations in at least seven Northern Hemisphere proxies (Fig. **4.11**). The majority of the coefficients of band-pass correlation (Table **4.4**) are quite significant. The peak-to-through amplitude of this periodicity could reach 0.2° C, and thus the cycle could have contributed substantially to the warming of the first part of the twentieth century. Likely sources of a hemispheric-scale century-long cycle are: (a) natural variability in the ocean-atmosphere system, (b) changes in solar activity, and (c) changes in activity of volcanoes (see more details in the Section 5.4). Solar origin seems less probable, since Ogurtsov *et al.* [384] did not find significant linear correlation between band-passed millennial solar reconstructions and reconstructions of hemispheric-scale temperature. Such a rhythm is, however, weakly manifested in the recent temperature multi-proxy of Mann *et al.* [301]. Much further research is needed to clarify the origins and character of the century-scale variation of temperature in the Northern Hemisphere.

4.8. Remaining Problems

An enduring problem in palaeoclimatic studies is the shortness of the time period usually available for calibration - often no more than the last 90 to 100 years. This is not enough to assess the quality of reconstruction of long-term (periods of hundreds of years and more) temperature changes. Moreover, even within this short interval the ARS effect complicates the calibration process. Thus the ability of palaeoclimatology to reconstruct the long-term (low frequency) and probably the most powerful climate fluctuations raises questions. Further consideration of millennia-length temperature proxies strengthens the doubts. Actually, visual

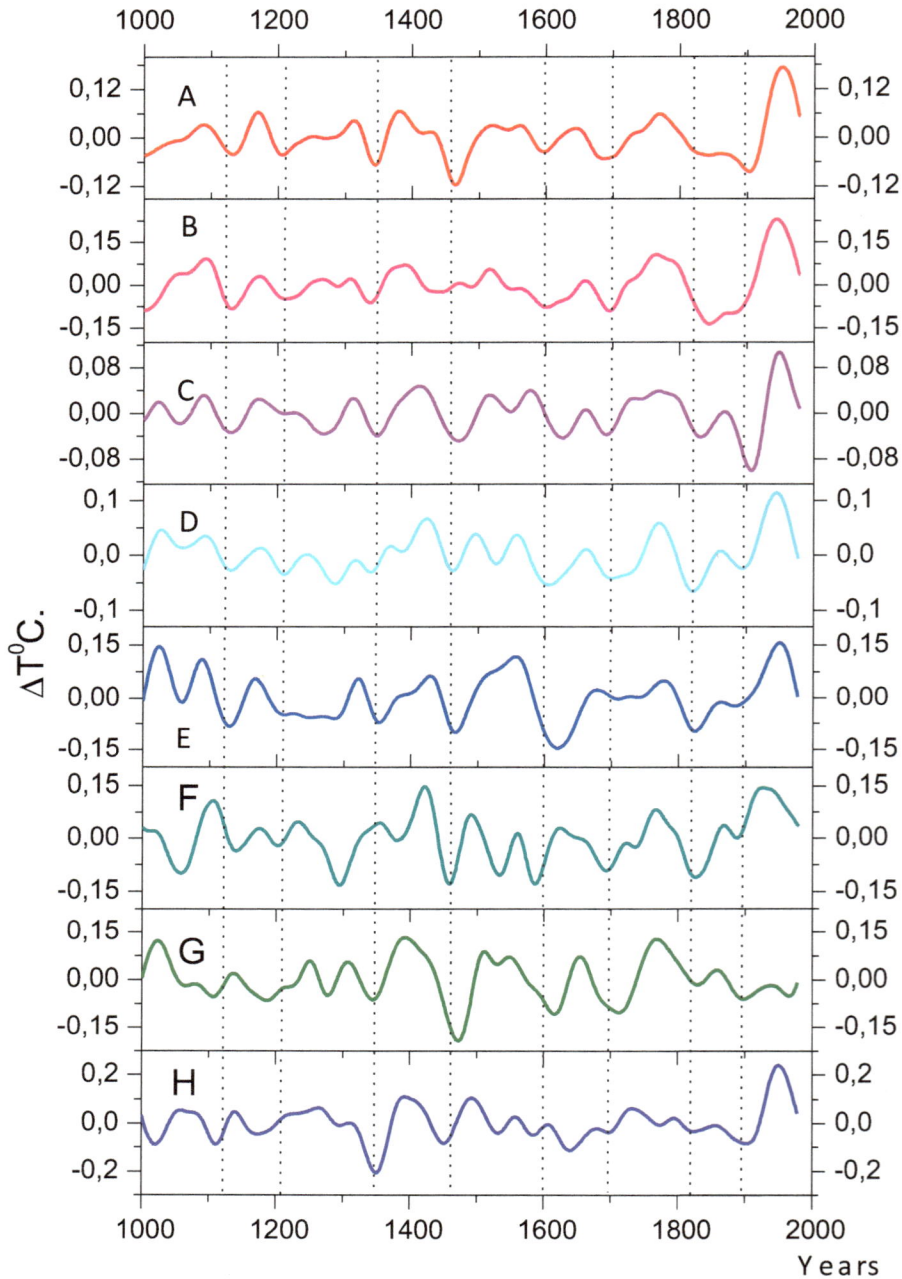

Fig. (4.10). Temperature proxies wavelet-filtered in the 55- 147-year scale band: (**A**) Mann *et al.* [359]; (**B**) Jones *et al.* [299]; (**C**) Crowley and Lowery [362]; (**D**) Briffa [296]; (**E**) Esper *et al.* [298]; (**F**) Moberg *et al.* [363]; (**G**) Loehle [364]; (**H**) Mann *et al.* [301].

Table 4.4. Bandpass (55-147 year) correlation between different temperature proxies from AD 1000-1900. The significance of every correlation coefficient was evaluated by means of the statistical experiment described in Section 3.3. Red figures correspond to confidence level (c.l.) >0.98, italic magenta figures - c.l. 0.95 to 0.98, blue figures - c.l. 0.85 to 0.95, italic black figures - c.l. <0.85.

	NHJ	NHC	NHB	NHE	NHMb	NHL	NHM08
NHM99	0.67	0.57	*0.49*	*0.45*	*0.20*	0.68	*0.21*
NHJ		0.51	*0.50*	0.37	*0.22*	*0.41*	*0.21*
NHC			0.55	0.59	*0.47*	0.54	0.31
NHB				0.66	0.64	0.54	0.25
NHE					*0.19*	*0.43*	*0.15*
NHMb						0.35	*0.10*
NHL							0.30

inspection (see Fig. **4.6**) shows that the plotted reconstructions (individually) demonstrate clearly different patterns of temperature variability during the last 1000 to 2000 years. Wavelet spectra of the temperature reconstructions [385] show the differences even more obviously (Fig. **4.11**).

Both wavelet analysis and visual examination of the data have let to divide temperature reconstructions in the following groups [385, 386]:

(a)   Multi-proxy reconstructions NHM, NHJ and NHC (see Fig. **4.6A-C**) show an evident linear decrease of average temperature (up to 0.15-0.30°C) across the pre-industrial epoch (AD 1000 to 1880) and a sharp increase thereafter (the so-called hockey stick form). Temperature in the middle of the twentieth century is clearly highest over the 1000 years. NHM99, NHJ, and NHC reconstructions can be combined as a group, depicting a temperature pattern with a sharp rising shape - a hockey stick (IHS).

(b)   Reconstructions NHMb and NHL (Fig. **4.6F**, **G**) also demonstrate a linear trend in AD 1000 to 1880 but the temperature rise during the 20th century is not sharp. The warming of the twentieth century is comparable with that during the Medieval Warm Period (AD 800 to 1100), *i.e.* it is not unusual. A strong cyclicity with a period longer than 1000 years dominates in the spectra of these time series (Fig. **4.11F**, **G**). We further term the NHMb and NHL reconstructions millennial-variability (MV) reconstructions.

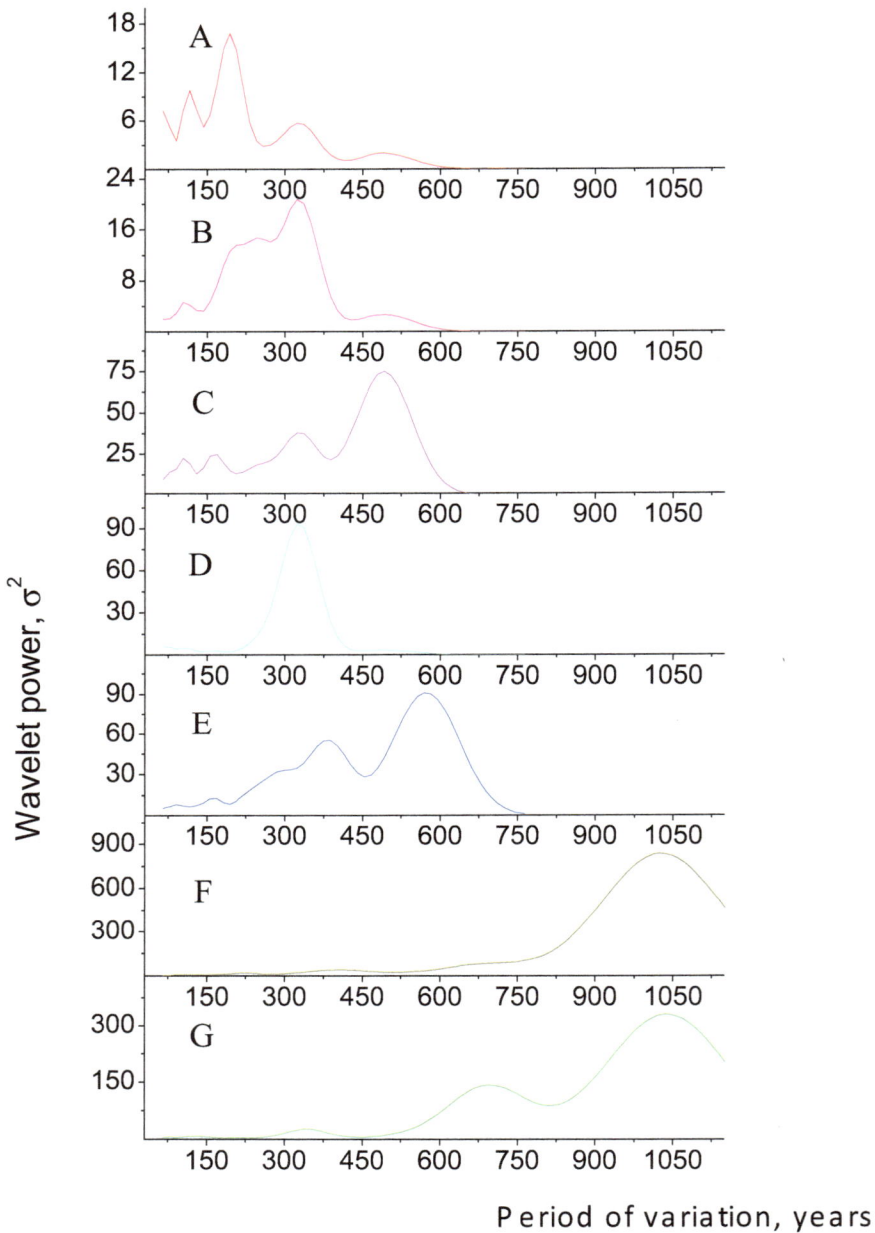

Fig. (4.11). Global wavelet spectra of: (**A**) Mann *et al.* [359]; (**B**) Jones *et al.* [299]; (**C**) Crowley and Lowery [362]; (**D**) Briffa [296]; (**E**) Esper *et al.* [298]; (**F**) Moberg *et al.* [363]; (**G**) Loehle [364]; (**H**) Mann *et al.* [301]. Linear trends were subtracted from series NHM99, NHJ and NHC prior to analysis.

(c)   Reconstructions NHE and NHB (Fig. **4.6D**, **E**) do not demonstrate a linear trend. Instead, their low-frequency variability is dominated by multi-centennial cyclicities (Fig. **4.11D**, **E**). The 20th century was warm, but the rise of temperature was not abnormal. We can combine NHE and NHB temperature records as a group showing the shape of multi-centennial variability (MCV).

The divergence between different temperature histories cannot be completely explained by differences either in different spatial coverage or in the standardization techniques [386]. Thus the reconstructions of IHS, MV and MCV kinds can be considered as three different clusters of seven temperature reconstructions. Bürger [387] analysed ten temperature paleoreconstructions using a more complicated technique and arrived at conclusion that they form five clusters, all of which are significantly mutually inconsistent. Use of the popular "spaghetti diagram" approach allows demonstration of discrepancies between different temperature reconstructions (Fig. **4.12**).

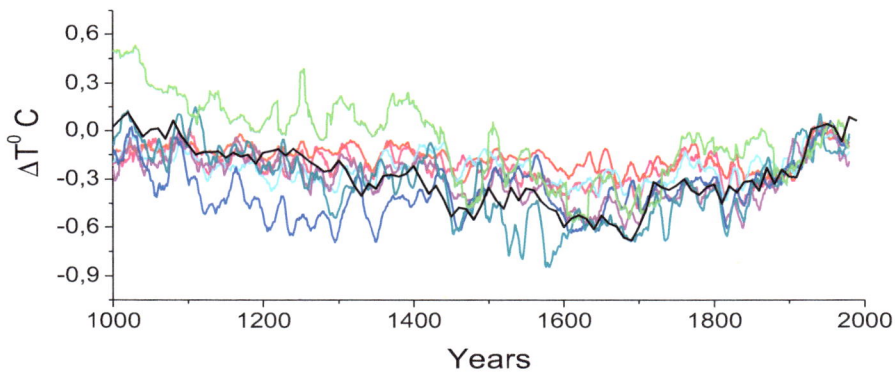

Fig. (4.12). Temperature reconstructions smoothed by 11 years. Red line - Mann *et al.* [359], pink line - Jones *et al.* [299], magenta line - Crowley and Lowery [362], cyan line - Briffa [296], blue line - Esper *et al.* [298], dark cyan line - Moberg *et al.* [363], green line - Loehle [364], black line - Mann *et al.* [301].

Evidently the available millennial-length palaeo-proxies contain some information about such important climatic episodes as the Medieval Warm Period (MWP) and the Little Ice Age (LIA) (Fig. **4.12**). Lamb [388] claimed that MWP and LIA were periods of global-scale negative and positive temperature anomalies, which occurred between AD 900 to 1300 and AD 1500 to 1700, respectively. Since then some doubts about the global character of LIA have appeared [389], but currently the amount of evidence for a global (or at least Northern Hemispheric) extent of the MWP and the LIA has increased [363, 390, 391]. Using the temperature reconstructions (Fig. **4.11**), however, it is practically impossible to determine the timing and amplitude of the MWP and the LIA.

The discussion about the divergence between global temperature reconstructions is connected with the problem of confidence and reliability of the available temperature reconstructions, a topic which became especially popular after the works of McIntyre and McKitrick [392-395]. The two Canadian specialists in mathematical statistics applied the methods of data processing used by Mann *et al.* [396] to the same source data and obtained a Northern Hemisphere temperature reconstruction quite different from the NHM record (see Fig. **4.13**). McIntyre and McKitrick [392] showed that the 20th century was warmest throughout the last 500 years but the 15th century was much warmer. McIntyre and McKitrick [392, 393] excited the scientific community and generated debate [397-400].

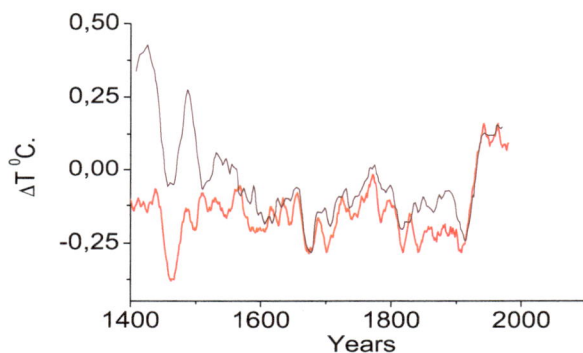

Fig. (4.13). The multi-proxy temperature reconstructions of Mann *et al.* [396] (red line), and McIntyre and McKitrick [392] (brown line), obtained using the identical initial data. Both records are averaged over 13 years.

The discussion continues, but it can be most logically concluded that substantial differences between the reconstructions of global temperature obtained by using the same initial data show that, despite evident achievements, the methods and approaches of palaeoclimatology still leave appreciable space for subjectivity. The question arises - what are the real features of temperature that can be captured by the reconstructions? In other words, if the available temperature reconstructions have appreciable differences, what are their common features? Bürger [387] has concluded that the temperature reconstructions disagree so much that there is no way to draw meaningful conclusions from them. On the other hand, Ogurtsov *et al.* [365] demonstrated, however, that in spite of their differences the reconstructions of the Northern Hemisphere temperature have at least one important common feature - the existence of a roughly regular century-scale cyclicity with a period of 50 to 130 years throughout the last 1000 years (see Section 4.7.3). Another common feature is apparent temperature increase during the 20th century, clearly manifested in all the proxies (Fig. **4.12**). The Intergovernmental Panel on Climate Change (IPCC) suggests that during the second half of the twentieth century the average Northern Hemisphere temperatures were likely was the warmest for the last 1,300 years [401]. This conclusion is based on a comparison of the temperature reconstructions with the instrumental record (see Fig. **6.10** of IPCC [339]). This approach may not be fully justified because of the ARS effect. Actually, if we assume that during the last two or three decades tree growth has ceased to reflect continuing warming, then we cannot be sure that this effect did not also take place in the past. That means that it is not easy to detect past climate epochs warmer than those Taking place during the calibration period using tree-growth-based proxies. In order to avoid this problem Ogurtsov *et al.* [365] used a new temperature reconstruction for the Northern Hemisphere obtained by Wilson *et al.* [402], who utilized 15 tree-ring proxy records that show no divergence effects for the last decades. Using this divergence-free time series Ogurtsov *et al.* [365] corrected eight millennial-scale temperature reconstructions of the Northern Hemisphere for the ARS effect and reanalysed them. They showed that, disregarding the work of McIntyre and McKitrick [392], there is a higher than 70% probability that in the extratropical part of the Northern Hemisphere the 1988-2008 time interval was the warmest two decades of the last 1000 years. Ogurtsov *et al.* [365] noted, however, that the applied procedure of correction for ARS effect includes some rather subjective assumptions and needs further improvement.

Thus, despite significant achievements and hard work in the field of palaeoclimatology over the last decades, our knowledge about the evolution of climate during the past one to two millennia is still far from complete. The available tree-ring proxies and multi-proxies of the Northern Hemisphere temperature represent different and sometimes conflicting histories of the long-term changes over the last millennium. They have only two evident common features: (a) presence of a century-long cyclicity, and (b) a considerable temperature increase during the 20th century. The available information even about such climatic extremes as the Little Ice Age and Medieval Warm period is incomplete. Thus, no range of dates is universally accepted as a time limit for both MWP and LIA. In addition, the amplitude of global temperature change during these two climatic episodes is also not well-known. Information about climate variability during the last 1000 years in the Southern Hemisphere is particularly poor because of the lack of palaeoclimatic records. It is very likely that currently we have more reliable knowledge about the past of solar activity than about the past of global climate, at least for the last millennium.

Several difficulties in reconciling the orbital climatic theory with observations still remain. First of all, calculations show that the 100,000-year eccentricity variation has a substantially smaller influence on solar forcing than precession or obliquity (40,000 and 23,000-year periodicity) and consequently might be expected to produce minor climatic effects. During the last 430,000 years, however, the strongest climate variability is associated with the 100,000-year cycle. Moreover, according to Milankovitch's theory, changes of climate in the Southern Hemisphere should correspond to significant changes in the Northern Hemisphere, and in some cases it is not so [403].

Increasing the spatial coverage of the Earth's surface by individual palaeo-proxies is one of the most important directions for further development of palaeoclimatology regarding the past millennium. Correspondingly, increasing the temporal resolution and age control of the Holocene proxy data is of major importance for palaeoclimatology of longer scales (lower frequencies). Moreover, the methods of reconstructing the full spectrum of past temperature variability from proxies need improvement as well. Further progress in paleoclimatology is important also for helioclimatic research, since the analysis of long proxies makes it possible to study long-term solar-climatic relationship.

CHAPTER 5

Possible Mechanisms of Solar-Climate Links

Abstract: Modern concepts of the physical mechanisms providing a potential link between solar as well as cosmic factors and climate are described. Possible climatic effects of solar luminosity, fluxes of galactic and solar cosmic rays, UV and microwave radiation are considered. A survey of the most important experimental evidence of the reality of solar-cosmic influence on climate is presented.

Keywords: Cosmic rays, solar-climatic relationship, solar irradiance.

Most life forms on Earth depend on solar energy. The existence or absence of a connection between solar activity and the Earth's climate has been claimed in many studies and thus the possibility of a solar contribution to global climatic variability is actively discussed currently. Without a doubt foreseeable progress in our present understanding of the changes of climate through the most recent past as well as the upcoming decades are of huge value for the future of humankind. Thus, studies of climatic responses to different natural and anthropogenic factors are crucially important. Understanding the physical mechanism involved in the reaction of the Earth's climate system to solar-terrestrial connections is one of the central topics of heliogeophysics. It is of particular significance at present since the solar-induced climatic change is a possible contributor to global warming.

5.1. Climatic System and Radiative Forcing

The climate system is an interactive system consisting of five main reservoirs: (a) the atmosphere, (b) the hydrosphere (ocean, rivers, and lakes), (c) the land surface (soil and vegetation), (d) the cryosphere (ice sheets, glaciers, permafrost and sea ice), and (e) the biosphere. The relationship between the integral heat capacities of the ocean, surface, and atmosphere is 77:0.5:1.1 [367]. Thus, the ocean is the most inertial part of the climatic system, responsible for the longest climatic variations. The climatic system changes due to:

(a)   natural unforced variability (internal oscillations of the climate system);

(b)   natural forced variability (variations in the energy output of the Sun, volcanic aerosols, *etc.*);

(c)   anthropogenic forced variability (changes in greenhouse gas concentrations, industrial aerosols, *etc.*).

Maxim Ogurtsov, Risto Jalkanen, Markus Lindholm and Svetlana Veretenenko

The principal law of climate is that the Earth is in a radiative equilibrium, that is, in a state of equality between the net energy coming in from the Sun and the outgoing thermal radiation. The energy balance model is represented by the well-known equation, in which the left part describes the incoming short-wave radiation and the right part the outgoing long-wave radiation:

$$\frac{S_0 \cdot (1 - \alpha)}{4} = \varepsilon \cdot \sigma \cdot T_e^4, \tag{5.1}$$

where: $\sigma = 5.7 \times 10^{-8}$ W m^{-2} K^{-4} is the Stefan-Boltzmann constant; $S_0 = 1367 \times$ W m^{-2} is the solar constant; $S_0/4 = I_0 = 341$ W m^{-2} is the global average insolation. It is the solar constant divided by four because the Earth gains heat from a disk that intercepts sunlight (with an area of πR^2) and loses heat from a sphere (with an area of $4\pi R^2$); α is the global planetary albedo. It determines the fraction of incoming solar energy reflected back to space (Table **5.1**); $I_0(1 - \alpha) = 239$ W $\times$ m^{-2} is the solar power available to the Earth's climatic system; ε is the emissivity of the surface. Emissivity is a property of the material that relates its actual radiative properties to that of an ideal black *body*. A black body has an emissivity $\varepsilon = 1.0$ and represents the maximum radiative ability (Table **5.1**); T_e is the Earth's *effective emission temperature* (K).

Table 5.1. Radiative properties of different surfaces according to Henderson-Sellers and Robinson [404], Sellers [405], Shipper *et al.* [406], and Kirkby [407].

Type of Surface	Albedo α	Infrared Emissivity ε
Tropical rainforests	~ 0.13	0.99
Deciduous forests	0.05-0.14	0.90-0.98
Natural grassland	0.14-0.20	0.90-0.95
Semi-deserts	0.24	0.92
Deserts	0.25-0.37	0.80-0.92
Seawater (calm, high zenith angle)	< 0.10	0.92-0.96
Sea ice	0.25-0.60	0.90-0.96
Snow	0.20-0.95	0.82-0.99
Low clouds (optically thick)	~ 0.40	0.68
Medium clouds	~ 0.40	0.55
High clouds (optically thin)	~ 0.40	0.33

Using the values of global planetary albedo $\alpha \cong 0.30$ and global emissivity $\varepsilon = 0.945$ [408] we obtain the effective temperature:

$$T_e = \sqrt[4]{\frac{S_0 \cdot (1-\alpha)}{4 \cdot \varepsilon \cdot \sigma}} \cong 258 \ K ,$$ **(5.2)**

or -14.8 °C. Thus, if the Earth had no atmosphere (or had an infrared (IR)-transparent atmosphere) it would be a frozen snowball. However, the atmosphere and clouds are not transparent to IR radiation. The majority (75%) of sunlight belongs to the visible and short IR ranges (0.40-1.5 μm). The atmosphere is quite transparent over these wavelengths, and thus at least 75% of solar radiation reaches the terrestrial surface. The heated surface of the Earth and the bottom troposphere emit radiation of a long IR range (8-30 μm) with a maximum of emission at $\lambda = 12$ μm. The atmosphere is appreciably opaque over these wavelengths. A considerable part of the escaping long-wave radiation is absorbed by clouds and greenhouse gases (mainly H_2O, CO_2, CH_4). This causes additional heating of the corresponding atmospheric layers. As a consequence they begin to emanate extra-long-wave radiation, some part of which is re-radiated out into space while another part comes back to the surface of the Earth. This reduces the Earth's by up to 0.615. Therefore greenhouse gases and clouds serve as an IR radiation shield and warm the Earth considerably. Indeed, using formula (5.2) with $\varepsilon = 0.615$ we obtain the actual surface temperature $T_s = 14.7$ °C. The difference $T_s - T_s \cong 30$ °C is a result of the *greenhouse effect*. Warming of 21 °C is provided by H_2O (vapour and clouds) and to 7 °C by CO_2.

Each physical factor which changes the equilibrium of incoming and outgoing energy in the Earth atmosphere (5.1) is called *radiative forcing*. Radiative forcing (usually in W × m^{-2}) thus determines the perturbation in the radiation energy balance of Earth's climate system. Negative forcing tends to cool the surface whereas positive forcing tends to warm it. The Sun is a primary driver of the atmospheric processes. It has been recently found out that fluxes of energetic particles entering the atmosphere are closely connected with the Sun: the Sun's magnetic activity drives changes of GCR intensity and the Sun is a direct source of SCR particles. Therefore, the Sun can influence the climate through:

(a)   changes in the energy input into the Earth's atmosphere through variations in the solar constant (integrated luminosity);

(b)   changes in physical and chemical characteristics of the atmosphere, which determine its transparency, by electromagnetic and corpuscular radiation;

(c)   changes in cloud cover by electromagnetic and corpuscular radiation.

However, it is difficult to clarify the physical mechanism of this link, since the total energy contribution of the particles modulated by the Sun is very little as compared to the energy of atmospheric processes (Table **5.2**).

Table 5.2. **Characteristic energy of the general earthly manifestations of solar activity and atmospheric processes according to Vitinsky *et al.* [53], Borisenkov [342] Pudovkin and [409].**

Type of Energy	Power (Erg/Day)
Energy brought into the Earth's atmosphere by TSI	1.6×10^{29}
Energy brought into the Earth's magnetosphere by the fast solar wind	up to 4×10^{24}
Energy brought into the Earth's magnetosphere by solar wind	up to 3×10^{23}
Energy brought into the Earth's atmosphere by the flux of GCR with E > 0.1 GeV	4×10^{21}
Energy brought into the Earth's atmosphere by water vapour (latent heat)	4×10^{28}
Energy release by thunder clouds	up to 10^{27}
The power of cyclones (anticyclones)	up to 10^{25}

It is evident (Table **5.2**) that the energy brought into the Earth's magnetosphere and the atmosphere by a majority of solar factors is a few orders of magnitude less than the energy of meteorological processes. The energy brought by the very large solar flare of August 1972 is also small (ca. 10^{24} erg), while the energy release of a baric system (cyclone or anticyclone) can reach 3×10^{25} erg in a few days. Thus, among the solar factors only the TSI (total solar irradiance) can influence atmospheric processes directly. However, changes of TSI over time - the solar constant - are rather small.

5.2. Total Solar Irradiance and Climatic Change

Changes of TSI - the radiation at all wavelengths received from the Sun - is often considered as a manifestation of solar activity, which is of key importance for solar-climatic relations. Direct measurements of TSI performed outside the Earth's atmosphere started with the launch of satellite devices in 1978. Three satellite TSI composites were constructed by generalization of data obtained in different space experiments and by means of different radiometers: the PMOD, the ACRIM, and the IRMB composites (Fig. **5.1**).

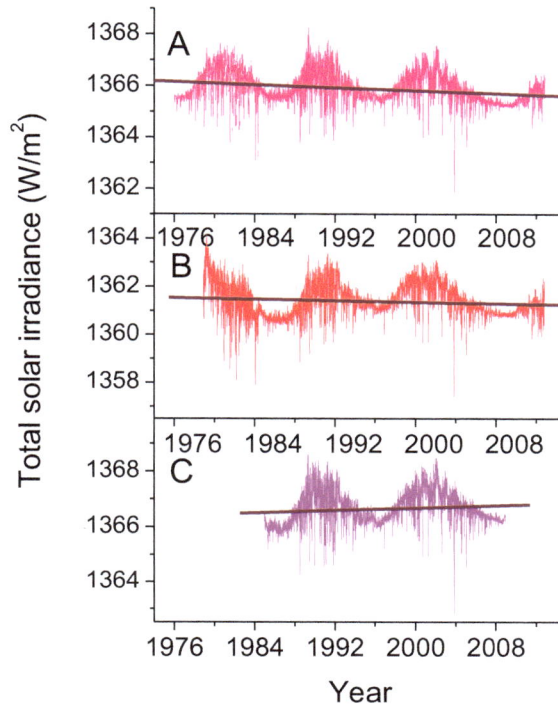

Fig. (5.1). Total solar irradiance records: (**A**) PMOD (Fröhlich [410], updated from ftp://ftp.pmodwrc.ch/pub/data/irradiance/composite/DataPlots/ext_composite_d41_62_1210a.dat); (**B**) ACRIM (Willson and Mordvinov [45], updated from http://www.acrim.com /RESULTS/ data/composite/composite acrim 1105.rtf); (**C**) IRMB (Dewitte *et al.* [411], updated from http://remotesensing.oma.be/meteo/view/ en/3385925-SARR.diarad.html). Thin curves - the raw data; thick lines - linear trends.

TSI clearly follows the general 11-year variation of solar activity (Fig. **5.1**). It is not reasonable to speak about any longer term variation because of disagreement between the data. For example PMOD composite shows a decline of the mean TSI value, which was ca. 0.5 W $\times$ m^{-2} in 1976-2012 (Fig. **5.1A**). Taking into account that the accuracy of long-term TSI measurement is 0.35 Wm^{-2} [412], this trend can hardly be assessed as statistically significant. The IRMB series has a weak rising trend in 1985-2008 and the ACRIM series has practically no trend in 1978-2012. Thus, an analysis of the three existing composite datasets does not allow us to draw any conclusion about the long-term (decades and longer) trend of TSI. Moreover, data obtained by the newest device, the Total Irradiance Monitor, have led the ACRIM collaboration to the conclusion that the real value of mean total solar irradiance is 4.6 W $\times$ m^{-2} lower than previously measured [413]. This indicates that the uncertainty connected to the degradation of the instruments over time and their calibration actually reaches a few W × m⁻². Such a large measurement uncertainty makes it almost impossible to find the existence of any secular variation in TSI during the last three cycles [414, 416].

A quasi 11-year cycle is the only periodicity whose presence in solar luminosity has been established absolutely reliably and unambiguously. However, this value is rather stable - the peak-to-trough amplitude of changes of TSI over the 11-year cycle is $\Delta S = 1.5\text{-}2.0$ W × m⁻² at the top of the atmosphere, that is, less than 1.5%. That is enough to provide a radiative forcing of $\Delta F = \dfrac{\Delta S\,(1-\alpha)}{4} = 0.26\text{-}0.35$ W × m⁻². For a rough quantitative estimation of the probable climatic reaction to this forcing we can multiply it by climatic sensitivity λ_c, the value which determines the reaction of the climate system to a special radiative forcing ΔF:

$$\Delta T = \lambda_c\,\Delta F \tag{5.3}$$

The sensitivity (usually in K × W⁻¹ × m²) may differ for different forcing and frequency bands (see Waple *et al.* [415]). This depends on a number of feedback processes which operate over different time scales - changes in atmospheric water vapour, cloud cover, snow-ice albedo, area of ice sheets, ocean circulation, and biogeochemical cycling. It should be noted that here as well as in the case of the radiocarbon exchange system, the response of ΔT to short-term fluctuations of ΔF is considerably attenuated by the thermal inertia of oceans, because of its large heat capacity and integrate variations in heat input. If we use for simplified evaluations (a) values of λ_c adopted by the IPCC [150] by means of analysis of the greenhouse effect of the last 100 years: $\lambda_c = 0.54\text{-}1.22$ W⁻¹ × m² × K, and (b) a damping factor of 0.25 for 11-year variation of ΔF [416], we find that an 11-year cycle in TSI provides a peak-to-trough variation in global temperature of 0.03-0.10 °C, which is insufficient for an effective contribution to climatic variability.

5.3. Energetic Cosmic Rays and Climatic Change

5.3.1. Ground-Based and Balloon Observations Testifying to a Connection Between Cosmic Ray Flux and Physical Properties of the Atmosphere

The Sun can affect the Earth's climate and weather not only directly *via* changes of luminosity variations of TSI but also indirectly through modulation of corpuscular radiation incoming into the atmosphere. Ney was the first who assume that cosmic ray flux could influence the weather [109]. This idea was further developed by Dickinson [110]. The problem is that the total energy input of the solar-modulated particles is much less than the energy of meteorological processes (see Table **5.1**). On the other hand, the effect of cosmic radiation on the atmosphere could be very nonlinear. Slight alterations of physical and chemical properties of the atmosphere, which do not need a lot of energy, can considerably alter its optical properties, change energy balance, and stimulate powerful climatic processes. Research on the possible connection between CR and atmospheric processes started in the 1960s. Recent decades have been a period of intensive investigation of solar activity and the effects of cosmic rays on weather and climate, which has produced many interesting results.

Let us list some of the more important results obtained by measurements performed at the surface and in the atmosphere:

1)   Schuurmans and Oort [417] revealed changes in the altitude of a number constant pressure levels (mainly 500 mb) in the lower stratosphere and troposphere after powerful solar flares. Their analysis covered the Northern Hemisphere north of 10° N and was based on a series of 81 thoroughly selected flares which occurred from July 1957 till December 1959. The duration of these effects was found to be 6-24 hours.

2)   Dmitriev and Lomakina [418] discovered an appreciable increase in cloudiness above four regions of the USA ($\varphi = 25\text{-}50°$ N, $\lambda = 65\text{-}130°$ W) one to two days after the solar X-ray flares.

3)   Kondratyev and Nikolsky [419, 420] investigated variations of the so-called meteorological solar constant - the integrated intensity of sunlight arriving in the troposphere (the solar constant weakened by absorption in the stratosphere). For this purpose balloon soundings of solar radiation over 25-33 km altitude were performed. The experiment took place during 1961-1970 in Volsk ($\varphi=52.0°$ N, $\lambda=47.4°$ E) and a total of 28 launches were executed. It was found that the meteorological solar constant reaches a maximum at $R_z = 50\text{-}70$ and drops down to 1.2% at $R_z > 70$. Kondratyev and Nikolsky [419, 420] ascribed this decrease to a reduction in atmospheric transparency due to the effect of solar cosmic rays.

4)   Pudovkin and Babushkina [421] investigated the variations of direct solar radiation flux measured under cloudless conditions onto a perpendicular plane over the two latitudinal belts of the USSR - the auroral ($\varphi \approx 67°$ N, five stations), sub-auroral ($\varphi \approx 60°$ N, three stations), and mid-latitude zones

($\varphi \approx 50°$ N, three stations). They found an appreciable (up to 10%) increase in solar insolation, which occurs one to four days after the beginning of intensive geomagnetic disturbances (Fig. **5.2**) and coincides with drops in GCR intensity (Forbush decreases).

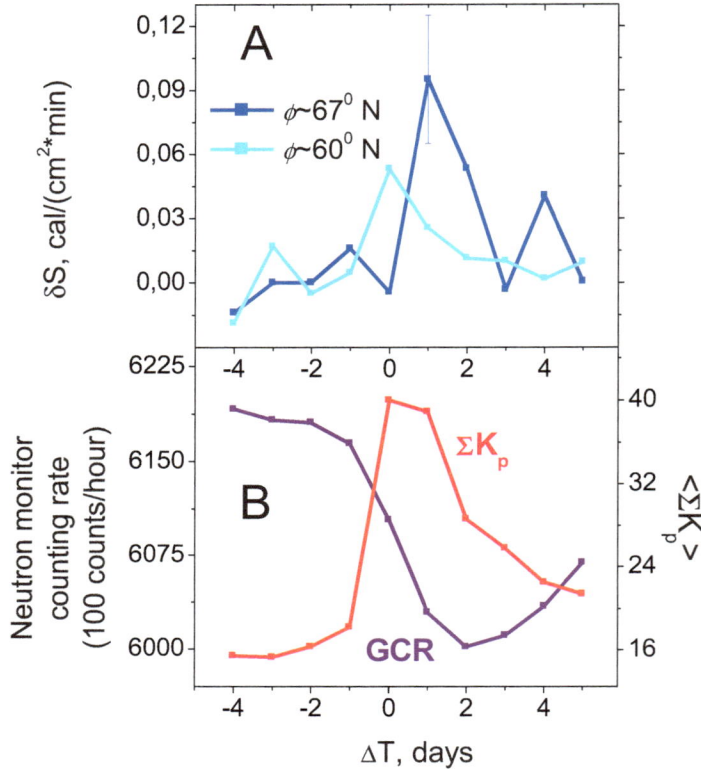

Fig. (5.2). Storm-time variations of: (**A**) the noon intensity of surface solar radiation in the auroral zone (blue curve) and the sub-auroral zone (cyan curve); (**B**) counting rate of Moscow neutron monitor (R_c=2.28 GV) (purple line) and daily geomagnetic index $\sum_8 K_p$ (red line).

Pudovkin and Babushkina [421] estimated the total amount of additional energy dissipated in the atmosphere as a result of an increase in the Blinova index of 10^{27} erg which may result in zonal circulation increase [426].

5)   Veretenenko and Pudovkin [422] studied connection between changes in atmospheric circulation characterized by the Blinova index and: (a) FDs through 1970-1982 (33 events), (b) SCR bursts at the Earth's orbit through 1980-1986 (56 events). The index of Blinova [423] is a parameter, which

characterizes zonal circulation of the lower atmosphere over middle latitudes. It is determined as $A = 10^3 \times \dfrac{\alpha}{\Omega}$, where α is the angular velocity of the zonal atmospheric flux averaged over the latitudinal belt $\varphi=45°\text{-}65°$ N and $\Omega=2\pi/86400$ is the angular velocity of the Earth rotation. They revealed statistically significant decreases of the Blinova index 1-4 days after the beginnings of FDs and SCR bursts accompanied by fluxes of high energy ($E_p> 90$ MeV) particles (27 events). The effects are illustrated with Fig. (**5.3**). The day zero ($\Delta t=0$) corresponds to the starting point of the events under study.

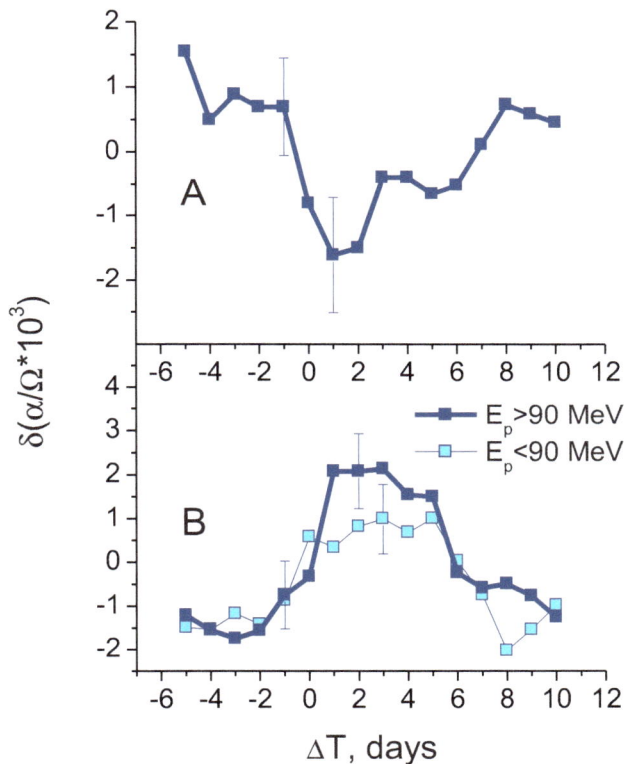

Fig. (5.3). The Blinova index variations at the 500 mbar pressure level associated with: (**A**) Forbush decreases of GCR; (**B**) solar proton bursts. Dark blue line - SCR events, accompanied by increases of particle fluxes with $E_p>90$ MeV (27 events). Blue line with cyan squares - SCR events, not accompanied by increases of particle fluxes with $E_p>90$ MeV (29 events). Figure after Veretenenko and Pudovkin [422].

The Blinova index is a parameter which characterizes zonal circulation of the lower atmosphere over middle latitudes. It is determined as $\delta = 10^3 \times \dfrac{\alpha}{\Omega}$, where α is the angular velocity of the zonal atmospheric flux averaged over the latitudinal belt $\varphi = 45\text{-}65^\circ$ N, and $\Omega = 2\pi/86400$ is the angular velocity of the Earth's rotation. At St. Petersburg ($\varphi = 60^\circ$ N) the value $\delta = 1.0$ corresponds to the linear wind velocity $v = 0.23$ m×s^{-1} [424, 425]. Changes in zonal atmospheric circulation velocity associated with FDs, and SCR pulses (Fig. **5.3**) can reach 0.4-0.6 m × s^{-1}. Pudovkin and Babushkina [421, 422] estimated the increase of the atmosphere's energy necessary for the observed intensification of circulation as 5-7 × 10^{26} ergs (power ≈ 10^{26} erg/day).

6)   Roldugin and Vashenyuk [427] studied variations of atmospheric transparency above Murmansk ($\varphi = 69^\circ$ N), Archangelsk ($\varphi=65^\circ$ N), and Leningrad (St. Petersburg, $\varphi = 60^\circ$ N) after six solar proton events. They found that one to three days after strong SCR events (30 April 1976, 3 April 1979, and 17 August 1979) the transparency at the wavelength $\lambda = 344$ nm decreased by a factor of 1.5-4.0. In two cases (8 April 1978 and 8 September 1979) transparency increased and in one case (11 May 1978) there was no apparent effect. The authors speculated that enhancement of atmospheric turbidity, detected in three cases, could be a result of the generation of additional aerosol layers under the action of multiple ions produced by solar protons.

7)   Veretenenko and Pudovkin [428] analysed changes of total cloudiness during 1969-1986 using the data of the USSR actinometrical network over three latitudinal areas: the auroral zone ($\varphi = 65\text{-}68^\circ$ N, four stations), sub-auroral zone ($\varphi = 60\text{-}64^\circ$ N, six stations), and middle-latitude zone ($\varphi = 50^\circ$ N, four stations). They observed a statistically significant decrease of total cloudiness one to two days after the beginnings of Forbush decreases over high-latitude belts. The cloudiness variations were most clearly pronounced in the sub-auroral belt. The effect disappeared at middle latitudes. Veretenenko and Pudovkin [428] concluded that the upper layer cloudiness (cirrus clouds) contributes mainly to the observed variations. Their result was further confirmed by Todd and Kniveton [429], who examined the satellite-based cloud data for 32 FDs (1983-2000) and found an appreciable reduction of high level clouds over the Antarctic plateau.

8)   Starkov and Roldugin [430] investigated changes of atmospheric
      transparency in spectral ranges of 344 and 627 nm over Murmansk
      (ϕ= 69°), Leningrad (ϕ= 60°), and Feodosia (ϕ= 45°) five days before and
      ten days after 101 geomagnetic perturbations in 1973-1976 and 1978. They
      found a reduction of transparency by 2-8% two days prior to the beginning
      of perturbation at both wavelengths. The amplitude of transparency
      variations decreases with a decrease of latitude. Starkov and Roldugin [430]
      linked the observed increases of atmospheric opacity to X-ray bursts.
      According to their estimations the 6% transparency variation results in a
      variation of incoming energy of 10^{27} erg/day.

9)   Veretenenko and Pudovkin [431, 432] studied the relationship between solar
      radiation input to the troposphere at middle and high latitudes and different
      helio/geophysical phenomena (solar flares, GCR intensity, auroral activity)
      using data of the network of Russian actinometric stations (1961-1986). The
      analysis was performed for three latitudinal belts: φ = 65-68° (four stations),
      φ = 60-64° (six stations), and φ = 50° (four stations). They found that the
      effects of GCR variations and solar flares on the half-yearly sums total
      radiation (sum of both direct and scattered radiation) are strongly dependent
      on latitude, with a positive correlation being observed in the mid-latitudinal
      zone and a negative one at higher latitudes. The change of the correlation
      sign happens at the latitude near 57° N. Veretenenko and Pudovkin [432]
      showed that in the auroral latitudinal belt the total radiation sum in the
      warm half-year period can change by 4-6% (±10 W$\times$m^{-2}) in the course of an
      11-year cycle of GCR flux (see Fig. **5.4B**).

      During the cold half-year period the connection between GCR intensity and
      solar radiation input is weaker (see Fig. **5.4A**). The effects revealed by
      Veretenenko and Pudovkin [431, 432], are likely to result from latitudinal
      dependence of cloud cover changes caused by SA/CR factors

10) Vanhellemont *et al.* [433] studied long-term stratospheric aerosol climatology
      including the data derived from atmospheric turbidity and transmission
      measurements during 1953-1974 and worldwide lidar observations through
      1974-1984. They found an evident linear relationship between the cosmic ray
      intensity (Climax data) and stratospheric aerosol optical thickness in the zonal
      band of 40-50° N for the period of 1953-1980. Vanhellemont *et al.* [433]
      showed that during this time interval changes in the cosmic ray signal resulted
      in a significant change in the aerosol signal two months later.

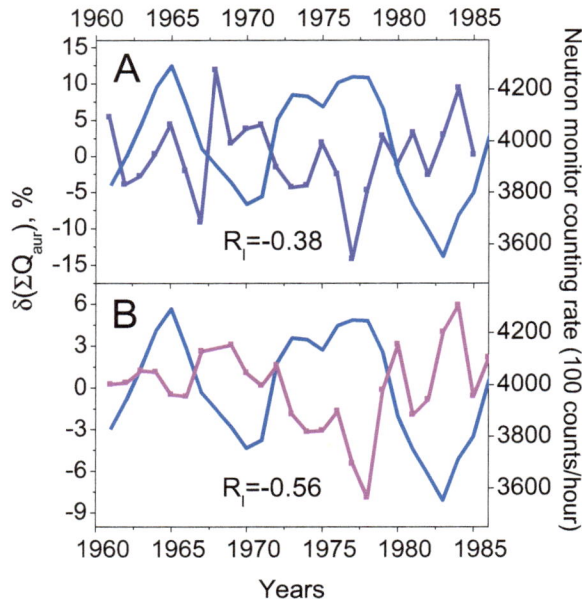

Fig. (5.4). Variations of the total radiation half-year totals $\delta(\sum Q_{aur})$ in the auroral ($\varphi \cong 65\text{-}68^\circ$ N) belt: (**A**) in the cold half-year period (violet line); (**B**) in the warm half-year period (magenta line). The blue line is a counting rate of the Climax neutron monitor (yearly average). (adopted from Veretenenko and Pudovkin [431, 432]).

11)   Harrison and Stephenson [434] examined data on diffuse solar radiation measured over the UK territory by surface weather stations since 1947. They compared the ratio of diffuse to total solar irradiance (cloud cover proxy) with the daily mean neutron count rate measured at Climax neutron monitor. They found a statistically significant link between diffuse radiation and cosmic rays and concluded that GCR influences cloudiness across the UK.

12)   Kancirova and Kudela [435] found a weak positive correlation over time scales longer than 1 year between cloud cover and cosmic ray intensity observed at Lomnicky stit mountain (49.40°N, 20.22° E, 2634 m above sea level).

The above studies as well as many others (*e.g.* [436, 437]) testify to the reality of the influence of solar and cosmophysical factors on weather and climate. The majority of these works describe a negative correlation between surface radiation and intensity of corpuscular and electromagnetic solar-cosmic radiation. That is, an increase of the flux of ionizing particles in the Earth's atmosphere causes: (a) a decrease of the atmospheric transparency (rise of optical thickness), and (b) an increase of cloud cover. Case (a) is an expression of the optical mechanism of the Sun-climate connection, and case (b) is a manifestation of the cloud mechanism of such a link. In both cases the amount of solar radiation reaching the lower troposphere decreases. Hereafter we will call this effect, indicating a negative correlation between cosmic ray intensity and surface radiation, CR-SR↑↓. It is important that changes in surface solar irradiance connected with the CR-SR↑↓ effect can reach 10^{26}-10^{27} erg/day, which is enough for the modulation of meteorological processes even at global scale.

Many researchers consider an enhancement of aerosol and cloud droplets due to increases of atmospheric ionization caused by fluxes of energetic particles as the most plausible physical mechanism of the effect. It should be noted that the results of some other authors contradict the aforesaid ones. For example, Roldugin and Tinsley [438] investigated variations of atmospheric transparency in 1978-1989 using data from 45 stations situated in the former USSR territory north of 55° N. They found that during periods of low stratospheric aerosol loading, atmospheric transparency decreases after Forbush decreases. Roldugin and Starkov [439] analysed data of eight Soviet stations ($\varphi > 55°$) and revealed a significant negative correlation between atmospheric transparency and Wolf number at wavelengths $\lambda = 344$ and 369 nm during 1972-1989. This result evidently contradicts the results of Kondratyev and Nikolsky [419, 420] and Veretenenko and Pudovkin [428, 431]. Possible causes of such discrepancy still need to be clarified.

Variations of the pressure field in the troposphere as well as changes in atmospheric circulation connected with intrusions of solar protons have also been observed [417, 426, 437]. We will call this effect CR-AC. Perturbations of atmospheric circulation could result in variations of cloud cover.

An increase in the concentration of atmospheric aerosols caused by fluxes of high energy solar protons in the atmosphere of the Earth has been observed experimentally a few times.

1) A group of Russian researchers headed by O.I. Shumilov detected a substantial rise in the aerosol backscatter ratio at Verhnetulomsk observatory ($\varphi = 68°$ N, $\lambda = 32°$ E) in February 1984 by means of lidar [440, 441]. Lidar measurements at the wavelength $\lambda = 694.3$ nm showed that a few days after the strong GLE event of 16 February 1984 the concentration

of aerosol particles with a diameter of more than 0.69 μm increased considerably (up to 50%) over altitudes of 15-25 km and, probably, more than 30 km (see Fig. **2** of [441]). The formed aerosol layer settled quickly (1-2 km/day) during 18-20 February.

2)   Marichev *et al.* [442] registered the generation of additional aerosol layers over altitudes of 40-45 km during geomagnetic storms in March of 1988 and 1989 by means of the lidar complex in Tomsk ($\varphi = 56°$ N, $\lambda = 85°$ E). The new aerosol layers settled with a velocity of 5 km/day. These authors connected the observed effect with the arrival of SCR particles and noted its irregularity.

3)   Mironova and Pudovkin [443] reported the occurrence of a new aerosol layer at Garmisch-Partenkirchen ($\varphi = 47°$ N, $\lambda = 11°$ E) at 10-12 km two days after the flares of 27 January 2002 and 24 August 2002.

4)   The formation of aerosol layers at 10-12 km height during the flare of 12 September 2000 was observed in Barcelona ($\varphi = 41°$ N, $\lambda = 2°$ W) and Hamburg ($\varphi = 53°$ N, $\lambda = 10°$ E) by means of lidar [444].

5)   The occurrence of a strong aerosol layer in the atmosphere of Arctic regions was registered by four lidar stations on 16 November 2000 [445]. According to the observation data this layer appeared at an altitude of about 38 km and consisted of particles with diameters of 30-50 nm. It existed until 12 February 2001, gradually falling downwards to 26 km. It is possible to assume a connection of this aerosol cloud with the powerful solar flare of 14 July 2000 and another big flare of 8 November 2000.

According to Kasatkina *et al.* [446], considerable enhancement of aerosol layers took place after the GLE events of 21-24 May 1990. However, after the GLE events of 2 May 1998, 14 July 2000, and 15 April 2001, no increases in aerosol concentration were observed. In addition, ground-based experimental observations testifying to the presence of a connection between solar activity, fluxes of ionizing particles, and surface radiation show instability of this relationship. The effect is sometimes present and sometimes absent. Moreover its sign can change. The situation is quite typical for helioclimatology. Lack of statistics could be one reason for this state of affairs. Indeed, all of the authors described above have worked with short and highly localized datasets covering only limited areas of the terrestrial surface. Satellite observations are usually free of this shortcoming. Spacecraft-borne monitoring of various types of atmospheric parameters began in the 1960s. Thus, time series of satellite observations of 30 years length and longer are currently available.

5.3.2. Space-Borne Observations Testifying to a Relation Between Cosmic Ray Intensity and Physical Properties of the Atmosphere

Along with satellite observations, new evidence has appeared for a CR-SR↑↓ mechanism and a link between solar activity and clouds. An enhancement of atmospheric aerosols due to high-energy cosmic rays has been reported in the following works:

1) Vanhellenmont *et al.* [447] analysed the data of stratospheric aerosol number densities obtained from the data of space-borne optical instruments of SAGE II (the Stratospheric Aerosol and Gas Experiment II) on-board the Earth Radiation Budget Satellite from October 1984 to March 2000. SAGE II observations used in the work covered an area $\varphi = 80°$ S to $80°$ N. Vanhellenmont *et al.* [447] compared the dataset with monthly means of neutron monitor counts measured at the Climax station. They revealed a causal relationship between the two data sets, showing that cosmic rays affect stratospheric aerosols in the stratosphere *via* a physical process with a residence time of a few months. The mode radius of aerosols detected by SAGE II instruments is 0.26-0.40 μm [448].

2) Veretenenko *et al.* [449] studied the responses of the polar atmosphere to a number of solar proton events of 15, 16, and 17 January and GLE of 20 January 2005. The corresponding chain of geomagnetic disturbances caused an increase in the daily sum of the K_p index up to 47.3 on 19 January 2005. Veretenenko *et al.* [449] analysed the data on stratospheric aerosol concentration measured by the GOMOS (Global Ozone Monitoring by Occultation of Stars) instrument installed onboard the ENVISAT satellite. They reported a noticeable increase in aerosol concentration at ca. $72°$ N at 10-14 km during 18-24 January (see Fig. **5.5**). The result was confirmed by Mironova *et al.* [450], who used aerosol data of SAGE III ($\varphi = 66.0$-$72.8°$ N) and OSIRIS (Optical Spectrograph and Infrared Imaging System onboard the Odin satellite, $\varphi = 60$-$89°$ S). Mironova *et al.* [450] found a significant increase of aerosol concentration in the Arctic atmosphere at 10-24 km on 22-28 January 2005. An analogous although weaker effect was detected in the Antarctic atmosphere.

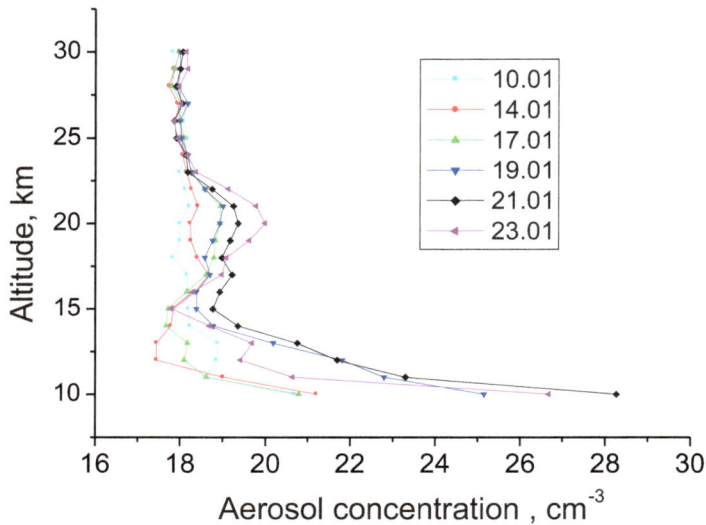

Fig. (5.5). Vertical profiles of aerosol concentration in the second part of January 2005 ($\varphi = 72°$ N, $\lambda = 180\text{-}100°$ W, and $20°$ W to $100°$ E) according to Veretenenko *et al.* [449].

3)   Mironova and Usoskin [451] analysed atmospheric reaction to a number of strong SPEs that took place during September-October 1989. They used aerosol data gained over high-latitude regions by two satellites with the space-borne optical instruments SAGE II ($\varphi > 60°\text{-}70°$ N) and SAM II (Stratospheric Aerosol Measurement II instrument aboard the Earth-orbiting Nimbus-7 spacecraft, $\varphi \approx 70°$ N, S). They revealed an appreciable increase in aerosol extinction at 10-16 km height in the Antarctic atmosphere a few days after the event of 29 September. The effective radius of aerosol particles also increased from about 0.25 to 0.5 μm. Mironova and Usoskin [451] concluded that a major SPE might result in generation of new particles and/or facilitate growth of pre-existing ultrafine particles in the polar stratosphere.

4)   New evidence of a connection between Forbush decreases and aerosol content and cloudiness was obtained by Svensmark *et al.* [452]. The authors studied the atmospheric response to 26 FDs in 1987-2007 by means of ground-based data of monitored aerosols in the atmosphere using solar photometers of the AERONET programme, performed by many stations well distributed over the Earth, and the space-borne data of:

(a)   the Special Sounder Microwave Imager, which survey variations in the cloud liquid water content over the Earth's oceans;

(b)   the Moderate Resolution Imaging Spectroradiometer on NASA's Terra and Aqua satellites (land and oceans), which measures the fraction of the liquid water in clouds; and

(c)   the International Satellite Cloud Climate Project (ISCCP) data on IR detection of low (< 3.2 km) clouds. The ISCCP system uses five geostationary and eleven polar-orbital satellites. This project, which started in 1983, provides the most complete database currently available. It contains data on three-hourly global coverage of clouds of different types and at various heights obtained by measurement of visible and IR radiances.

Svensmark *et al.* [452] revealed a substantial drop in all four values a few days after the FDs. The response peaked five to ten days after the GCR decrease began. The amplitude of the decrease in atmospheric parameters was found to be proportional to the FD strength. Svensmark *et al.* [452] considered this result as global-scale evidence of an obvious influence of the Sun's variability on aerosols and cloudiness. However, Calogovic *et al.* [453], who searched for changes in cloud cover after six of the largest FD events which happened in the period 1989 to 2001, arrived at the opposite conclusion. They reported that no reaction of global cloudiness to Forbush decreases at any latitude and height was found. The remaining controversy is likely a result of still poor statistics. This point is particularly important for analysis of an FD-meteorology relationship, because Forbush decreases are often accompanied by strong solar flares. The fluxes of high-energy SCR generated by flares that arrived to the Earth prior to FD can produce in the low atmosphere effects opposite to those caused by a drop in GCR intensity. Separation of these two types of effects is a complicated task which requires a lot of detailed and comprehensive information. (see Pudovkin and Veretenenko [426]).

5)   Very intriguing results were obtained by satellite observations of cloudiness. Analysis of the ISCCP data by Svensmark and Friis-Christensen [454] showed a strong positive correlation of the monthly time series of global cloudiness and galactic cosmic rays. This finding has generated intensive discussion [455-458]). As a result of this debate and further research it was concluded that only low (< 3.2 km) cloudiness correlates with GCR positively [459, 460]. Some authors have come to the conclusion that low cloudiness correlates better with solar irradiance than GCR flux [461]. In any case, space-borne cloud observations of ISCCP until the end of the

1990s brought strong evidence in favour of the reality of the CR-SR↑↓ effect over a time scale of 10-20 years Fig. (**5.6**). The uncertainty of monthly cloud data is ca. 2%.

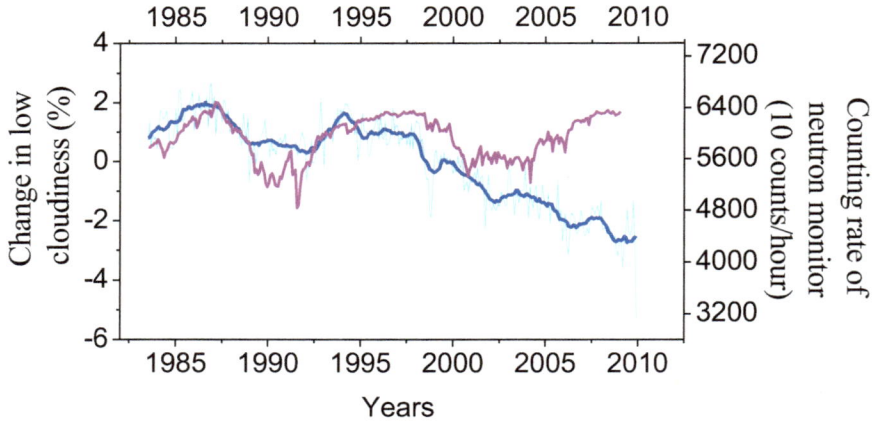

Fig. (5.6). Monthly averages of ISCCP-D2 global low cloudiness together with the data on GCR flux measured by the neutron monitor in Kiel. Cyan line - monthly data on the global average of low (<3.2 km) cloud cover anomalies (ftp://isccp.giss.nasa.gov/pub/data/D2CLOUDTYPES); thick blue line - annual averages. Magenta line - counting rate of the Kiel neutron monitor (http://cr0.izmiran.rssi.ru/kiel/main.html).

A considerable positive correlation between GCR and low cloudiness is clearly observed during 1983-2001. However, a decline of the correlation between galactic cosmic ray flux and cloudiness in the second part of the 1990s and a change of its sign after 2002 make any conclusion about a potential link between cosmic rays and clouds rather questionable [462, 463]. Moreover, Evan *et al.* [464] have reported that the ISCCP dataset is not suitable for the secular trend examination due to modifications in the satellite view angles. That is why Gray *et al.* [465] and Laken *et al.* [466] concluded that the currently available data provide no appreciable support to the hypothesized association between cloudiness and cosmic rays. On the other hand Usoskin [278] argued that the most appropriate GCR-cloud analysis should be applied to specific geographical regions. Such an analysis by Voiculescu and Usoskin [467] showed that over some key geographical areas the linkage between cloud cover and cosmic-ray-induced ionization and ultraviolet (UV) radiation is persistent during the time interval 1984-2009. The cloud response to solar-cosmic factors, however, has substantial geographical distribution: over some areas, cloudiness varies in phase with solar activity, while over other areas they are in anti-phase. Voiculescu and Usoskin [467] revealed that the positive correlation between low cloudiness and GCR is manifested most distinctly at: (a) the southern Atlantic ocean, (b) the western Indian ocean, and (c) the high-latitude North Atlantic ocean. The correlation is less distinct at (d) the southern Pacific Ocean and (e) continental East Asia, and is unclear over the Antarctic. A weak negative correlation between low cloud and GCR was found at the western coast of South America and a part of the Indian Ocean. Zones of significant correlation between GCR/UV and middle and high cloudiness were also found. The results of Voiculescu and Usoskin [467] thus give new evidence of a solar-climatic connection but also exhibit its complexity. Voiculescu and Usoskin [467] showed that the reaction of the climatic system to solar activity changes probably has a complicated spatial and altitude distribution.

Research on the response of atmospheric circulation to short-term changes in the flux of high-energy particles by Artamonova and Veretenenko [468] and Veretenenko and Thajll [469] confirms the geographical heterogeneity of SA/CR effects.

6)   Artamonova and Veretenenko [468] studied changes of the lower atmospheric pressure associated with 48 Forbush decreases which were not accompanied by intensive solar proton fluxes during 1980-2006. They used NCEP/NCAR re-analysis data. The NCEP/NCAR re-analysis archive is a comprehensive dataset created by the National Center for Environmental Prediction (NCEP) together with the National Center for Atmospheric Research (NCAR) in order to study atmospheric processes on the global and regional scales. This archive includes numerous meteorological parameters obtained from different sources (ground-based, aircraft and satellite measurements, balloon sounding, *etc.*). The Forbush events were selected for the cold period (October-March), because during half of the year temperature gradients in the troposphere are high and thus cyclonic activity is most intense. Before and after FDs, a noticeable pressure growth was revealed in the North and South Atlantic regions. Fig. (**5.7**) demonstrates variations in near-surface pressure before and after FDs. The maxima of pressure rise were observed on the third and fourth days after the event onsets over northern Europe and western Siberia in the Northern Hemisphere, as well as on the fourth and fifth days along the outer margin

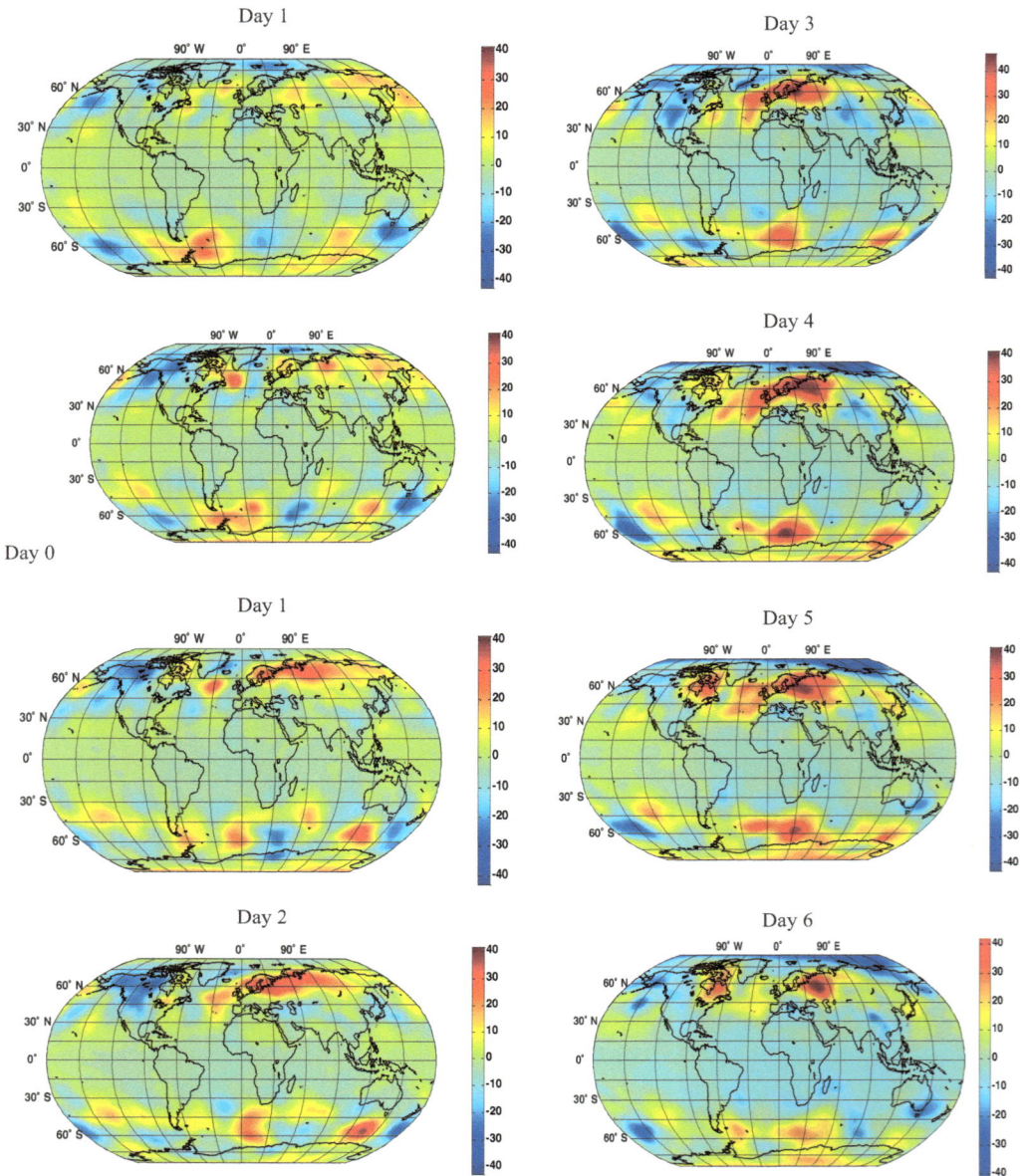

Fig. (5.7). Maps of the mean changes of geopotential heights for the 1000 hPa level (in geopotential metres) during 48 Forbush decreases for the period 1980-2006 (October-March) (after Artamonova and Veretenenko [471]).

of the Antarctic Ocean in the Southern Hemisphere. Both maxima are observed at middle latitudes, with $\varphi = 45\text{-}70°$ N and S, correspondingly.

Artamonova and Veretenenko [468] assumed that the observed pressure increase results from the intensification of anticyclones in these regions.

7)   Veretenenko and Thajll [469] analysed the effects of 48 solar proton events accompanied by fluxes of particles with energies higher than 90 MeV which occurred during 1980-1998 on the evolution of cyclonic activity at extra-tropical latitudes of the both hemispheres using NCEP/NCAR data. It was found that these SPEs were accompanied by an evident strengthening of cyclonic processes at middle latitudes, mainly over oceans. In the Northern Hemisphere the largest cyclone strengthening (significance of effect > 0.99) takes place in the North Atlantic near the south-eastern coast of Greenland and is due to the intensification of cyclone regeneration [470] (Fig. **5.7, 5.8**). In the Southern Hemisphere some cyclone intensification was found over the Antarctic Ocean near the coasts of the continent.

The areas susceptible to SPE influence are the regions of high temperature contrasts and low geomagnetic cutoff rigidities. The results of Artamonova and Veretenenko [468-471] and Veretenenko and Thajll [469] corroborate a complex latitude-regional pattern of atmospheric sensitivity to solar and cosmophysical factors.

8)   Veretenenko and Ogurtsov [472] demonstrated that the sign of SA/GCR effects on atmospheric pressure could be linked to a particular macro-circulation epoch.

They used the data of the NCEP/NCAR archive and analysed the spatial and temporal structure of the effects of the Sun's activity and GCR flux variations on the lower atmospheric circulation for two time intervals, 1953-1981 and 1982-2000. These two periods are characterized by different types of evolution of atmospheric macro-circulation. According to the system of Vangengeim [473] and Girs [474], atmospheric processes in the zone $\varphi = 35\text{-}80°$ N can be categorized into three fundamental types by direction of the air mass transfer: meridional (C), westerly (W), and easterly (E). Each of these forms is derived from the daily atmospheric pressure maps over the northern Atlantic-Eurasian region and is characterized by a specified distribution of troughs and ridges and the type of organization of the Rossby wave pattern. The duration of each circulation form (W, E, or C) through the year is expressed in days. Thus, the total yearly duration of all the three circulation forms is therefore 365. The main specific feature of the

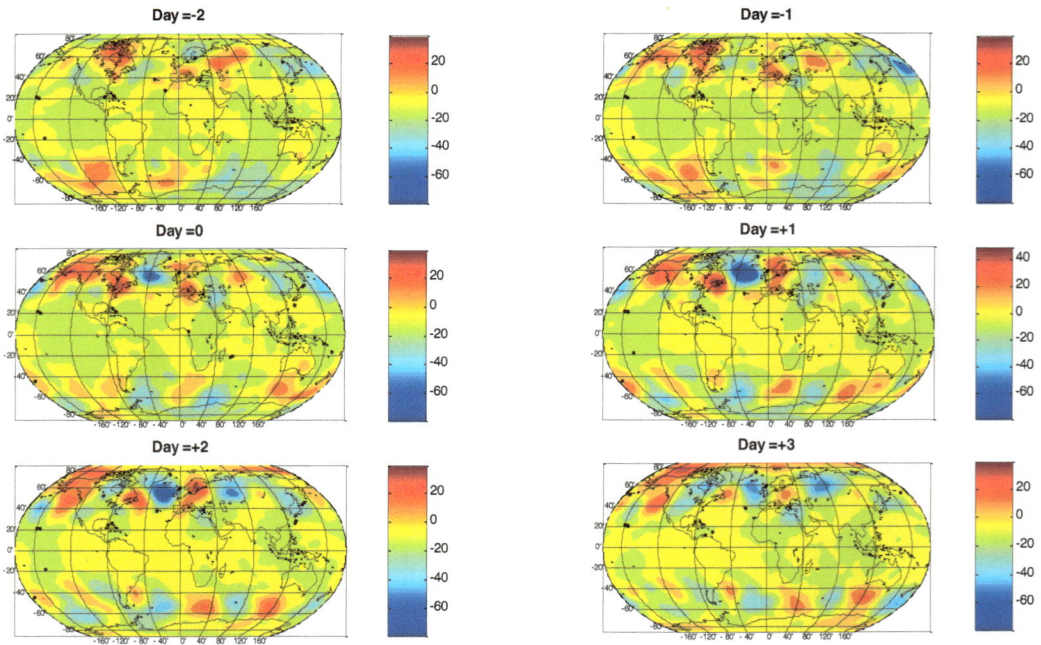

Fig. (5.8). Mean maps of the changes of geopotential heights for the 500 hPa level (in geopotential metres) during 48 solar proton events for the period 1980-1998 (October-March) (after Veretenenko and Thejll [465]).

western (zonal) form W is the development of west-east air transport with small-amplitude waves in the pressure field (Fig. **5.9**). The C form is characterized by the development of the meridional atmospheric circulation, when stationary or slowly moving large-amplitude waves are formed in the pressure field. The E form is similar to the C form, but it is distinguished by the opposite location of troughs and ridges.

During 1950-1981 the occurrence frequency of the meridional (E) forms increased, while zonal (W) circulation weakened (Fig. **5.9B**). The period of 1982-2000 has a sharp decrease of circulation E. At the same time, the recurrency of zonal circulation rises. Maps of the correlation coefficients between mean yearly values of geopotential heights of the 700 hPa isobaric level and the Climax neutron monitor counting rate for the periods 1953-1981 and 1982-2000 demonstrate that: (a) the regions of the most pronounced GCR-pressure correlation are the high-latitude sector of the Northern Hemisphere (positive correlation) and zones close to polar frontal

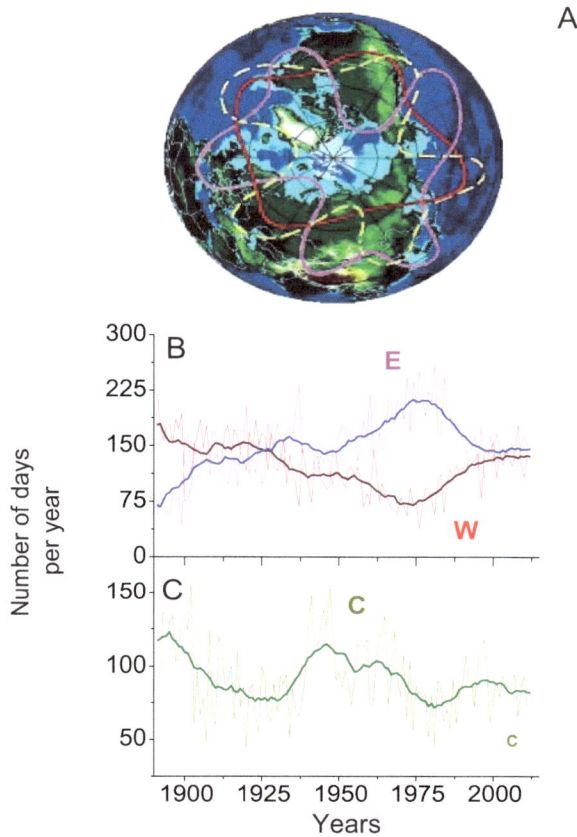

Fig. (5.9). (A) distribution of of ridges and troughs at 500 mb level for the circulation forms W (red line), **E** (magenta line) and **C** (yellow line); (**B, C**) annual frequencies of occurrence of the main forms of the large-scale circulation. The data were taken from Girs [474] and AARI [475].

zones (regions of sharp temperature contrasts) in both hemispheres; and (b) correlations during 1953-1981 and 1982-2000 have opposite signs (Fig. **5.10**). Veretenenko and Ogurtsov [472] hypothesized that the change of the sign of SA/GCR effects on atmospheric pressure variations in the early 1980s might be a result of the transformations of the large-scale atmospheric circulation.

Thus, Veretenenko and Ogurtsov [472] show that atmospheric reaction to SA/CR factors, in addition to having a spatial and a temporal structure, possibly depend on the state of the climatic system determined by its internal variability. This is in line with the results of Labitzke and van Loon

(a)

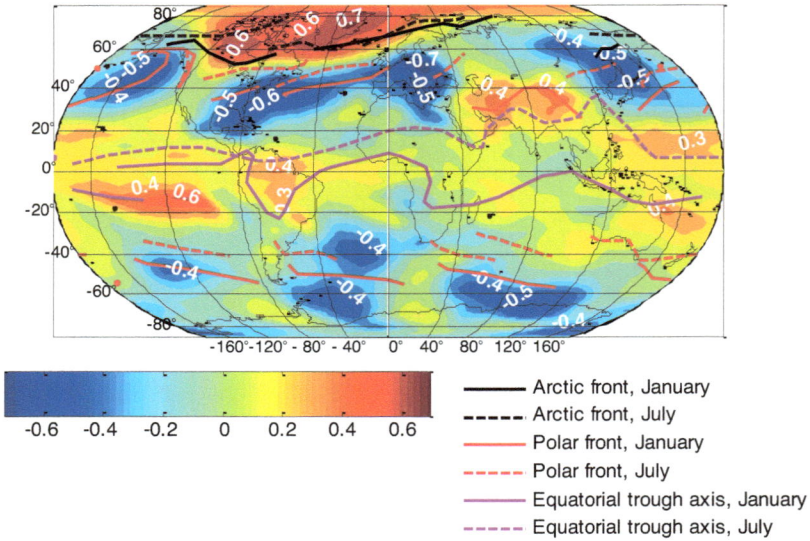

(b)

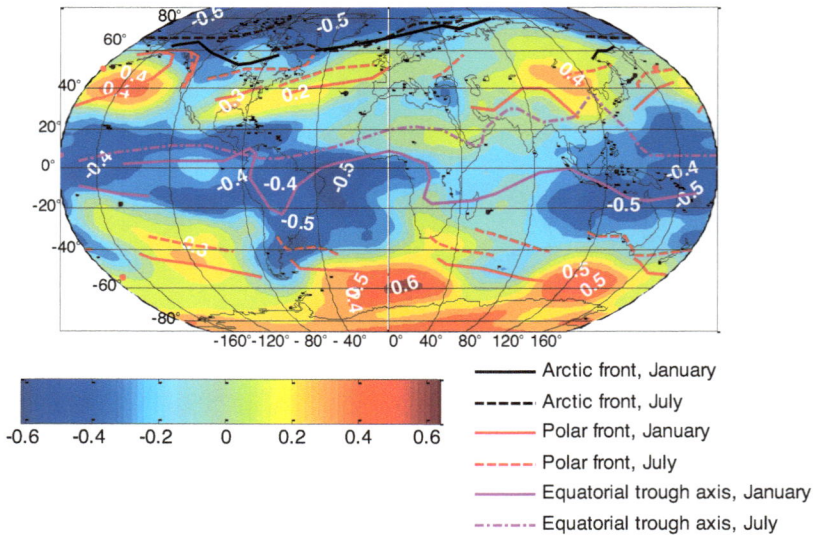

Fig. (5.10). Distribution of the correlation coefficients between mean yearly values of geopotential heights of the 700 hPa isobaric level and GCR flux intensity for: **(a)** 1982-2000 and **(b)** 1953-1981. The positions of the climatological fronts according to Khromov and Petrosyants [476].

[436], who have reported that the effect of the Sun's activity on temperatures in the stratosphere is different for the different phase of the quasi-biennial oscillation (QBO) - an approximately two-year natural oscillation of easterly and westerly zonal winds in the equatorial lower stratosphere.

9)   The most recent evidence of both regional distribution and temporal variability of a solar-climatic link was obtained by Maliniemi *et al*. [477]. The authors analysed the measurements of energetic (30-100 and 100-300 keV) electrons made by the NOAA/POES (National Oceanic and Atmospheric Administration/Polar Orbiting Environment Satellites) satellites, which have been operating since 1979. They found a statistically significant correlation between energetic electron precipitations and:

(a)   the index of North Atlantic Oscillation (NAO), a north-south seesaw variation in atmospheric mass, which is determined as the difference in atmospheric pressure at sea level between the Bermuda-Azores High and the Icelandic Low, and

(b)   surface air temperature in certain geographic regions of the Northern Hemisphere during 1980-2010. The strongest negative correlation was found over northeast Canada/Greenland and the strongest positive correlation in northeast Siberia. Maliniemi *et al*. [477] also found that the effect is considerably stronger and spatially wider through the easterly QBO phase. Maps of the atmospheric response to the influence of galactic cosmic rays [472] and energetic electrons (Fig. **6** of Maliniemi *et al*. [477]) has similar features. Thus a lot of empirical evidence of the reality of a relationship between the influence of cosmic rays and the physical state of the terrestrial atmosphere over different time scales has been obtained by means of both ground-based and satellite observations (Tables **5.3** and **5.4**). They show: (a) a negative correlation between the cosmic ray intensity and surface radiation (CR-SR↑↓), that is, a positive correlation between cosmic ray flux and cloud cover or/and aerosol content; (b) a connection between cosmic ray intensity and atmospheric circulation, which is usually rather complex. However, the effects are based on rather poor statistics and often unstable. Moreover, the results of some researchers (Roldugin and Tinsley [438], Calogovic *et al*. [453], and Erlykin and Wolfendale [463]) apparently contradict the proposed connection between cosmic rays and lower atmospheric processes.

Table 5.3. Empirical evidences for a connection between solar modulated cosmic radiation and the physical condition of the Earth's atmosphere over daily time scales. A means data of actinometrical (ground-based) stations, M data of weather (ground-based) stations, L data of lidar (ground-based) observations, R radiosonde data, and S satellite data.

Authors	Time Interval of Observation	Type of Data	The Revealed Effect	Geographic Coverage of Observation	Type of Effect
Tinsley and Deen [478]	1953-1985	M	Link between FDs and winter cyclone intensity	Northern Hemisphere	CR-AC
Pudovkin and Babushkina [421]	1961-1984	A	Increase of direct surface solar radiation during FDs particularly in auroral zones	USSR, $\varphi = 50\text{-}67°$ N	CR-SR ↑↓
Veretenenko and Pudovkin [422]	1970-1986	M	Intensification of atmospheric circulation after SCR events and reduction after FDs	USSR, $\varphi = 45\text{-}65°$ N	CR-AC
Roldugin and Vashenyuk [427]	1972, 1976, 1979	A	Decrease of atmospheric transparency 1-3 days after SCR events	Murmansk, 69° N Arkhangelsk, 65° N Leningrad, 60° N	CR-SR ↑↓
Shumilov *et al.* [440]	February 1983	L	Increase of stratospheric aerosol concentration at 15-25 km after GLE event	Verkhnetulomsk, $\varphi = 68°$ N	CR-SR ↑↓
Marichev *et al.* [442]	March 1988, March 1989	L	Increase of stratospheric aerosol concentration at 40-45 km after geomagnetic disturbances	Tomsk, $\varphi = 56°$ N	CR-SR ↑↓
Todd and Kniveton [429]	1983-2000	S	Reduction of high-level cloudiness after FDs	Antarctica	CR-SR ↑↓
Svensmark *et al.* [452]	1987-2007	S	Reduction in liquid cloudiness and concentration of fine aerosol particles after FDs	Globe	CR-SR ↑↓
Laken *et al.* [479]	1986-2006	S	Relationship between the rate of GCR flux change and the most fast mid-latitude cloud decreases	$\varphi = 30\text{-}60°$ N $\varphi = 30\text{-}60°$ S	CR-AC
Svensmark *et al.* [480]	2000-2006	S	Significant response of cloud fraction, effective emissivity, and liquid water content after FDs	Globe	CR-SR ↑↓
Veretenenko and Thajll [469]	1980-1998	M, R, S	Increase of cyclone activity over mid-latitude ocean after strong SPEs	Globe	CR-AC

Table 5.4. Empirical evidence for a connection between solar-modulated cosmic radiation and the physical condition of the Earth's atmosphere over inter-annual and decadal time scales.

Authors	Time Interval of Observation	Type of the Data	The Revealed Effect	Geographic Coverage of Observation	Type of Effect
Kondratyev and Nikolsky [420]	1962-1970	R	Decrease of atmospheric transparency above 25 km when Wolf number rises from 50-70 to 170	Volsk, φ =52° N	CR-SR ↑↓
Veretenenko and Pudovkin [431, 432]	1961-1986	A	Effect of GCR and solar flares on surface irradiance, particularly in auroral zone	φ = 50-68° N	CR-AC
ISCCP [459, 460]	1984-2004	S	Positive correlation between low-level cloudiness and GCR intensity	Globe	CR-SR ↑↓
Vanhellenmot *et al.* [433]	1953-1980	L, M, A	Positive correlation between stratospheric aerosol optical depth and GCR intensity	φ =40-50° N	CR-SR ↑↓
Veretenenko and Ogurtsov [472]	1953-2000	M, R, S	Complex (regional and temporal structure) response of atmospheric pressure to GCR variations	Globe	CR-AC
Maliniemi *et al.* [477]	1980-2010	S	Complex (regional structure) response of atmospheric circulation and surface air temperature to energetic electron precipitations	Northern Hemisphere	CR-AC

The next question is connected with the solar-related factor, which actually transfers solar influence to the lower atmosphere. The plausible mechanism of the CR-SR↑↓ effect is connected with the possible link between atmospheric ionization and concentration of aerosol particles, including cloud condensation nuclei (CCN) in the atmosphere. Pudovkin and Raspopov [481], Raspopov *et al.* [482], and Marsh and Svensmark [459] have proposed the following sequence of physical processes: activity of the Sun → fluxes of high-energy particles (SCR and GCR) → ion production in the atmosphere → formation of new ultra-fine particles in the atmosphere → concentration of aerosol particles in the stratosphere and CCN in the troposphere→ atmospheric transparency and cloud cover → surface solar irradiance → surface temperature. In the framework of this mechanism a so-called 'grey filter' exists in the atmosphere (in the stratosphere mainly). The filter blocks the penetration of sunlight to the surface of the Earth

[481] and the efficiency of this optical screen is directly connected to the flux of ionizing radiation. Another mechanism of a GCR-cloud connection has been proposed by Tinsley [483, 484] *via* the global atmospheric electric current. Electric current flows between the surface and ionosphere (h = 55-80 km). It has a density about 10^{-12} A $\times$ m^2 while the total atmospheric current is ca 1800 A [96]. The global circuit causes a downward vertical current in fair (non-thunderstorm) weather regions. This ionosphere-surface current density J_z passes through clouds, charging droplets and aerosols charging at their boundaries - zones, where potential sharp gradients in air conductivity can occur. Influence of the solar-induced changes in atmospheric ionization on the global circuit (both GCR and SCR effects) provides a possible link between clouds and variability of the Sun [483, 484]. Three potential mechanisms of the influence of the electric current on clouds have been proposed:

(a)   *Electroscavenging* is a result of electrically enhanced efficiency of the collision between charged particles with liquid droplets (*e.g.* Tinsley *et al.* [484]).

(b)   *Electrofreezing* is connected to the possible influence of electric properties of aerosol on the rate of freezing of thermodynamically unstable supercooled water droplets [478].

(c)   *Electrocoalescence* is connected to the possible increase in coalescence between charged droplets, which could influence the droplet size or number [485].

Electrofreezing - the electrical enhancement of the phase transition between supercooled water and ice - of thermodynamically unstable supercooled droplets in high-level clouds can be responsible for the CR-AC relationship as well. According to Tinsley and Deen [478], a downward vertical current in the troposphere facilitates electrisation of cloud droplets. The charged supercooled droplets freeze and form ice crystals more effectively. Thus electrofreezing increases the rate of formation of ice crystals, which are capable of settling at lower altitudes and glaciate in the mid-level clouds. The glaciation process releases latent heat that promotes cyclone intensification. So, a mechanism linking solar activity and atmospheric dynamics includes the following chain: activity of the Sun→ fluxes of high-energy particles (SCR and GCR) → ion production in the atmosphere → vertical current density J_z → charging of cloud droplets → electrofreezing of supercooled droplets→ latent heat release→ intensification of cyclones → atmospheric circulation.

Stozhkov *et al.* [486] proposed the idea that the high-energy cosmic rays effectively influence thundercloud activity. Stozhkov *et al.* [486] noted that ionization of the atmosphere by cosmic rays is one of the key factors required for the generation of thunderstorm electricity and lightning. Stozhkov *et al.* [95] further demonstrated a connection between short-term changes in cosmic ray flux and precipitation over the former USSR territory. A CR-thunderstorm link is another possibility for solar influence on atmospheric circulation.

The mechanisms described above were elaborated using the empirical evidence of a solar-atmosphere connection over time scales from a few days to 10-20 years. Palaeoclimatology makes it possible to examine this relationship over longer time scales.

5.4. Century-Scale Variation in Climate and Solar Activity

5.4.1 Century-Scale Variability in Palaeodata

It is important to examine whether the century-type hemispheric-scale climatic rhythm (see Section 4.7.3) is connected with the solar Gleissberg cycle. Correlations between century-scale cycles in sunspot number and Northern Hemisphere temperature throughout the pre-instrumental epoch show that century-long variations in solar and hemispheric-scale temperature proxies correlate weakly during the last (pre-instrumental) millennium (Table **5.5**). This can be considered as evidence of an absence of any physical linkage between the century-scale climatic variability and the solar Gleissberg cycle. The other possibility is that both century-long periodicities are linked but:

(a)   proxies of solar activity and climate used in the work do not record the temporal behaviour precisely enough and the possibilities for errors increase going back into the past;

(b)   the connection between century-type variations of the Sun's activity and hemispheric-scale temperatures has a nonlinear and complicated character. In that case the coefficient of linear correlation between them could be small even if the relationship is strong;

(c)   the century-scale climatic rhythm of solar origin is distorted by natural inherent oscillations of the climatic system.

Table 5.5. Bandpass (55-147 years) correlation between wavelet-filtered temperature and solar proxies during the period 1000-1890. Large red figures correspond to c.l. > 0.98, italic magenta figures to c.l. = 0.95-0.98, blue figures to c.l. = 0.85-0.95, and small black figures to c.l. < 0.85. Abbreviations for the Northern Hemisphere temperature reconstructions: NHM99 = Mann *et al.* [359], NHJ = Jones *et al.* [299], NHC = Crowley and Lowery [362], NHB = Briffa [296], NHE = Esper *et al.* [298], NHMb = Moberg *et al.* [363], and NHL = Loehle [364], and for the sunspot proxies: R_{SOL} = Solanki *et al.* [176], R_{NAG} = Nagovitsyn *et al.* [197], and R_{OG} = Ogurtsov [270].

	NHM99	NHJ	NHC	NHB	NHE	NHMb	NHL
R_{SOL}	0.12	0.08	0.06	0.15	−0.24	*0.42*	0.00
R_{NAG}	0.28	0.31	0.17	0.18	−0.19	0.40	0.08
R_{OG}	0.25	0.20	0.14	0.13	−0.29	**0.60**	0.08

De Jager and Usoskin [487] support the first assumption by revealing significant (c.l. > 0.97) correlations between the smoothed proxies NHM, NHJ, NHMb, NHB, and NHC and the group sunspot number over the last 350 years. De Jager and Usoskin [487] used a triangular smoothing function with a half width of nine years and thus selected variability at multi-decadal and longer scales. Their results testify that the use of more accurate instrumental data increases solar-climatic correlation. Some evidence of a non-linear character of the solar-climate link was obtained by Raspopov *et al.* [488] and White *et al.* [489]. It should be noted that over some regions, for example northern Fennoscandia, the linkage between century-long variations in SA and climate is much more distinct [120]. In addition, changing volcanic activity might be a natural climate-forming factor distorting solar influence. It is known that powerful volcanic eruptions accompanied by input of sulphate aerosol into the stratosphere can result in global cooling. It is evident (Fig. **5.11**) that a potential long-term (decades to centuries) warming of the Earth due to increasing activity of the Sun, at least during the nineteenth century, could be partly compensated by amplification of volcanic activity. Thus, careful assessment of possible volcanic effects on climate is an integral part of solar-climatic research.

5.4.2. Volcanic Forcing of Terrestrial Climate

The impact of volcanic eruptions on climate has been known for a long time. Volcanoes can eject both ash and gases like sulphur dioxide (SO_2) high into the Earth's atmosphere and thus decrease atmospheric transparency. But the lifetime of ash in the atmosphere is usually no more than a few months and therefore it

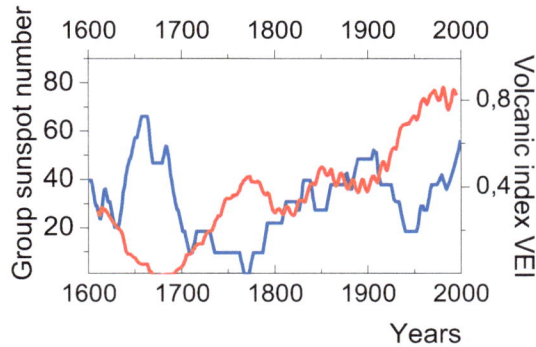

Fig. (5.11). Relations between volcanic explosivity index - blue line - and group sunspot number - red line - smoothed over 50 years. Volcanic data were taken from Zielinski *et al*. [490] and Briffa *et al*. [491].

cannot influence fluctuations of global climate much. On the other hand sulphur dioxide quickly becomes converted to sulphate aerosols, whose lifetime in the atmosphere can be much longer (Table **5.6**)].

Table 5.6. Lifetimes of aerosol particles in terrestrial atmosphere [492].

Atmospheric Reservoir	Altitude (km)	Lifetime
Lower troposphere	1-3	A few days
Upper troposphere	about 10	2-3 weeks
Lower stratosphere	10-20	9-12 months
Upper stratosphere	30-50	3-5 years
Lower mesosphere	50- 60	about 10 years

It is evident that aerosols in the troposphere are quickly washed out by precipitation (Table **5.6**). In the stratosphere, where precipitation is absent and the vertical motion of air is weak (its stability is due to a positive altitudinal gradient of temperature), sulphate aerosol (effectively scattering incident sunlight) can stay for several years, which is long enough to force changes of the Earth's climate on a global scale. Aerosol in the stratosphere can exist in the form of: (a) sulphate aerosol and (b) polar stratospheric clouds (see Section 5.6). Stratospheric sulfate aerosol is a suspension of particles with a size of 0.1-1.0 μm and consists mainly of 75% water solution of sulphuric acid (H_2SO_4). The liquid phase prevails up to heights of 17-20 km, and the solid (ice) phase generally does so above this level. A considerable part of stratospheric aerosol is usually concentrated in the Junge layer. Its thickness is about a few kilometres and the middle part is usually situated at 18-20 km altitude. Sulphate aerosol in the stratosphere is formed by condensation of the sulphuric acid vapour, which is generated from SO_2. The main source of SO_2 in the stratosphere is volcanic activity, with a minor contribution from anthropogenic emissions and OCS, produced in marine surface waters from photolysis of dissolved organic material [493]. Volcanic eruptions can be characterized by the volcanic explosivity index, which reflects the volume of ejected material on a logarithmic scale (Table **5.7**).

Table 5.7. General parameters of volcanic explosions. The tabular data on VEI and frequency of eruptions are mainly according to Chernavskaya and Cherenkova [494].

Sample of Explosion	VEI	Bulk Volume of the Eruption (m^3)	Mean Column Height (km)	Frequency of Eruptions	Mass of Sulphate Aerosol Delivered to Atmosphere (Mt)
Ruis, Columbia, 1985	3	10^7-10^8	12-19	Once per year	<0.1
Agung, Indonesia, 1963	4	10^8-10^9	19-26	Once per 2 years	0.1-5
El-Chichon, Mexico, 1982	5	10^9-10^{10}	26-33	Once per 10 years	5-20
Pinatubo, Philippines, 1991	6	10^{10}-10^{11}	33-40	Once per 10^2 years	20-110
Tambora, Indonesia, 1815	7	10^{11}-10^{12}	40-47	Once per 10^3 years	110-600
Toba, Indonesia, 70-75 kA BP	8	10^{12}-10^{13}	47-54	Once per 10^6 years	600-3000

The height of a column of emissions of volcanic products (ashes, tephra, SO_2) is calculated by means of the formula [495]:

$$H = 7.176 \times \log_{10}(M) - 60.521, \tag{5.4}$$

where H is the column height in kilometres and M is the total erupted mass in kilograms. The total mass of SO_2 transported into the stratosphere by eruption can be estimated by the formula of Pyle *et al.* [491]:

$$M = 10^{(0.75 \text{ VEI} - 0.21)}, \tag{5.5}$$

where M is the mass of SO_2 in kilotons. The magnitude-three eruptions can produce emission columns reaching the tropopause altitudes (15-17 km in the tropics, 10-12 km in the middle latitudes, and about 10 km at high latitudes) (Table **5.7**). However, as most volcanoes (> 70%) are located at low (< 40°) latitudes, that is, in the high-tropopause zone, only about 30% of eruptions with

VEI = 3 will inject material above the tropopause. Thus, they provide only minor loading of volcanic species [495].

Eruptions with VEI $\geq$ 4 are effective sources of stratospheric sulfur dioxide. The average interval between such eruptions is circa 1.8 years [495]. If the source of stratospheric aerosol is situated at extra-tropical latitudes, aerosol expands rather quickly over the corresponding hemisphere but penetrates into the other hemisphere more slowly. If the aerosol source is situated close to the equator, the aerosol expands over both hemispheres. In the stratosphere SO_2 enters the following chain of chemical reactions (see *e.g.* Eisel and Tanner [496]):

$$SO_2 + OH + M \rightarrow HSO_3 + M,$$

$$HSO_3 + O_2 \rightarrow SO_3 + HO_2, \qquad\qquad\qquad (5.6)$$

$$SO_3 + H_2O + M \rightarrow H_2SO_4 + M,$$

where M is molecules of atmospheric gases (N_2, O_2). As a result, the total mass of sulphate aerosol becomes ca. 2.5 times larger than the mass of its precursor, gaseous SO_2, transported into the stratosphere by volcanoes Zielinski [497]. Therefore the basic source of sulphuric acid in gas phase is SO_2, and the basic loss mechanism is condensation of the sulphuric vapour acid on already existing aerosol particles. Hence a steady-state approach yields as a first approximation:

$$K \times [SO_2] \times [OH] = [H_2SO_4]/t_s, \qquad\qquad\qquad (5.7)$$

where t_s is the lifetime of gaseous H_2SO_4 with respect to condensation (about 10^3 s in the lower stratosphere), and $K = 1.5 \times 10^{-12}$ cm$^3 \times$s^{-1} is the rate of reaction with OH radicals. Usually the time interval between the injection of sulfur dioxide in the stratosphere and its full transformation into sulphate aerosol (lifetime of SO_2 in the stratosphere) is 30-40 days, but sometimes it can reach 100 days. The single-scattering albedo of sulphate aerosol is close to 1.0 when the particle radius is less than 1 μm. This means that the majority of aerosol particles in the stratosphere scatter sunlight and do not absorb it. The backscattering coefficient for aerosol particles is 0.17-0.25 (Kondratyev and Cracknell [498]; IPCC [150]) thus about one fifth of the incident solar radiation is reflected upwards to space. On the other hand, particles of sulphate aerosol absorb IR radiation; that is, the stratospheric aerosol layer has some greenhouse effect, too. In principle, aerosol affects climate probably both seasonally and regionally. Calculations by Luther [499] have shown that stratospheric aerosol cools the atmosphere below 23 km over the areas with albedo $\alpha > 0.35$ and warms it over areas with $\alpha < 0.35$. However, aerosol warming is negligible on the global scale and the net effect of the added aerosol in the stratosphere from volcanic eruptions is to decrease the sunlight at the surface and cool the Earth. Model calculations performed by Toon and Pollack [500] have shown that at $\lambda = 0.55$ μm the background (equilibrium) optical depth of the stratosphere is $\tau_{str} \cong 0.005$. Experimental estimations of the average optical thickness of unperturbed stratosphere (periods 1960-1963 and 1972-1975 when there were no strong volcanic eruptions) give similar values of 0.002-0.005 [501]. This is low as compared to the optical depth of the troposphere $\tau_{trop}(\lambda = 0.55$ μm$) = 0.125$ [501]. However, volcanic eruptions can increase stratospheric optical thickness considerably. Stothers [502] examined the optical depth of the stratosphere after the Tambora eruption using the data of astronomy, glaciology, and history. He obtained the value of $\tau_{str} = 1.3$ in September 1815. Sato *et al.* [503] estimated a value of $\tau_{str} = 0.2$ after the Krakatau eruptions from the ice-core acidity data. It is evident that powerful eruptions decrease the atmospheric transparency significantly. The mean globally averaged optical depth of the stratospheric aerosol layer can be evaluated by means of the formula provided by Stothers [502]:

$$\tau_{str} = 6.510^{-15} \times M, \tag{5.8}$$

where M is the mass of the sulphate aerosol in grams. Radiative forcing of sulphate aerosol can be assessed by the formula of Charlson *et al.* [504]:

$$\Delta F = D \times S_0 \times T^2 \times (1 - A_c) \times (1 - \alpha)^2 \times \beta \times \tau, \tag{5.9}$$

where D is the fractional day length (0.5), S_0 is the total solar irradiance, T is atmospheric transparency above the aerosol layer (the part the solar radiation that is transmitted by the atmosphere above the layer of aerosol), A_c is the fraction of sky covered with clouds (usually close to 0.6), α is the mean albedo of the surface, β is the backscattering coefficient, and τ is the areal mean optical depth of aerosol. Global radiative forcing can be roughly estimated by the formula of Emile-Geay *et al.* [505]:

$$\Delta F = \Delta F_k \times (M/M_k)^{2/3} \tag{5.10}$$

where $\Delta F_k = 3.7$ W $\times$ m^{-2} is the forcing of the Krakatau eruption and $M_\kappa = 50$ Mt is the mass of sulphate aerosol ejected by Krakatau. It should be noted that conversion between aerosol optical depth and radiative forcing is still uncertain. Thus evaluations made using formulas (5.9) and (5.10) can differ considerably, particularly at high optical thicknesses. Even estimations of radiative forcing of the Pinatubo eruption studied by modern science range from 2.25 W $\times$ m^{-2} [506] to 4.7 W $\times$ m^{-2} [507] (Table **5.8**).

Table 5.8. Characteristics of four powerful historical volcanic eruptions and their impact on the stratosphere. Data according to Rampino and Self [508], Zielinski [497], and Tahira *et al.* [509]); optical depths and radiative forcing were calculated using formulas (5.8) and (5.10) respectively.

Volcano	The Date of Eruption	VEI	Energy of Eruption (Mt of TNT)	Volume of Emitted Material (km^3)	Mass of Sulphate Aerosol Burden into the Stratosphere (Mt)	Optical Depth of the Aerosol Layer τ	Radiative Forcing ΔF (W$\times$m^{-2})
Pinatubo	June 1991	6	70	10	15-20	0.10-0.13	1.6-2.0
Krakatau	August 1883	6	200	15-20	22-50	0.14-0.32	2.1-3.7
Tambora	April 1815	7	A few thousands	200	28-200	0.18-1.30	2.5-9.3
Toba	71 $\pm$ 5 kA BP	8	about 10^5	2500	1000-2000	Up to 13	Up to 43

The duration of elevated turbidity depends on the energy of eruption: aerosol delivered to high altitudes lives longer in the atmosphere (see Table **5.6**). Usually aerosol of volcanic origin is removed from the stratosphere after three years, but in the case of the most powerful eruptions (injection of sulfur in the upper layers), it can stay in the atmosphere for five to eight years [495]. The relationship between the optical depth of global volcanic aerosol and global mean temperature anomaly is not known precisely either. Harris and Highwood [510] used the formula:

$$\Delta T = 11.3 \times (1\text{-exp}(-0.164 \times \tau_{str})) \tag{5.11}$$

where ΔT is in degrees Celsius. Sigurdsson [511] estimated the amplitude of global cooling as:

$$\Delta T = 5.9 \times 10^{-5} \times S^{0.31}, \tag{5.12}$$

where S is the amount of sulfur released by the eruption in grams ($\cong$ 1/5 of the total sulphate aerosol mass).

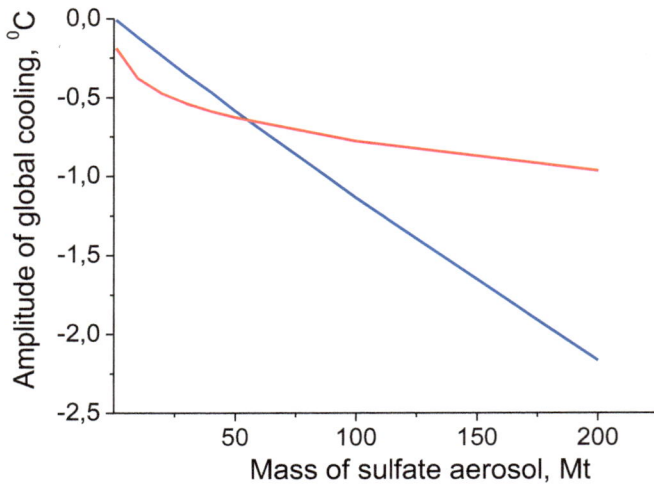

Fig. (5.12). The response of global temperature to volcanic aerosol injection into the stratosphere. Red line - formula of Sigurdsson [511]; blue line - formula of Harris and Highwood [510].

Thus, volcanic eruptions undoubtedly influence short-term (of the order of several years) climatic fluctuation, although the character of this communication is likely complicated and nonlinear (Fig. **5.12**). Briffa *et al.* [312], Shumilov *et al.* [512] and McCarroll *et al.* [361] have obtained clear evidence of volcanic effect on global temperature at longer time scales. Zielinski *et al.* [513] argued that the Toba mega-eruption pushed the globe into a cool period that lasted 200 years. Huang *et al.* [514] have concluded that the eruption of Toba caused cooling of the surface of the South China Sea by 1 °C during 1000 years.

Most likely a powerful volcanic eruption (or a cluster of eruptions occurring during a short time interval) is able to amplify the feedback mechanisms, whose action considerably extends the corresponding climatic episode. It is obvious that if volcanic activity indeed causes decadal-to-centennial climatic variations, the corresponding change in temperature can obscure the century-scale periodicity of solar origin.

5.4.3. Century-Scale Variability in Instrumental Data: Climatic Changes and Length of Solar Cycle

It has been demonstrated that long-term ($T > 30$ years) periodicity in global temperature resembles the corresponding cyclicity in sunspot numbers since the mid-19th century, which might be considered as a manifestation a solar-climatic link (see Fig. **5.13**). However, the smoothed temperature here leads solar activity, thus violating the cause-effect relationship. Friis-Christensen and Lassen [62] have reported that the problem could be solved using solar cycle length (SCL) as an index of the Sun's activity instead of sunspot numbers; SCL leads sunspot number by one to three cycles [54, 515]. Although the statement of Friis-Christensen and Lassen [62] has generated active debates [516, 517], the existence of such a link has been confirmed for regional temperatures [97, 518, 519]. On the other hand, physical processes accountable for this connection have not been clarified so far, which calls for new research efforts to search for alternative explanations.

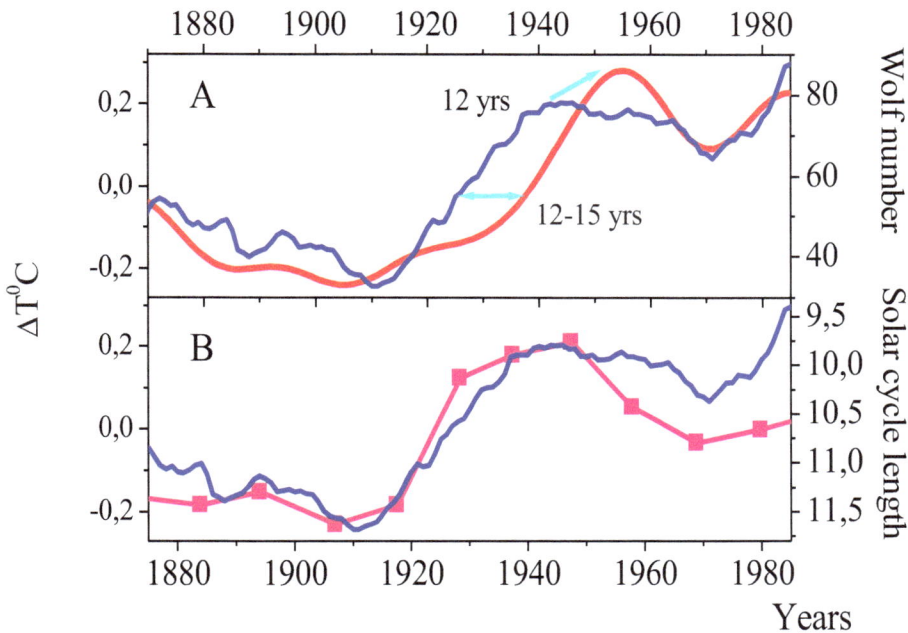

Fig. (5.13). Annual temperature of the Northern Hemisphere [315] smoothed by 15 years - violet curve and (**A**) Wolf numbers smoothed by the Fourier-filter (frequencies above 0.04 years^{-1} are reduced) - red curve; (**B**) the length of a solar cycle smoothed on three points - pink curve.

Ogurtsov *et al*. [264] have demonstrated a negative correlation between century-scale periodicities in nitrate abundance in Greenland ice and sunspot numbers. It was shown that the Gleissberg cycle in NO_3^- concentration leads the respective cycle in Wolf number by 15-18 years during the last three centuries (Fig. **5.14**).

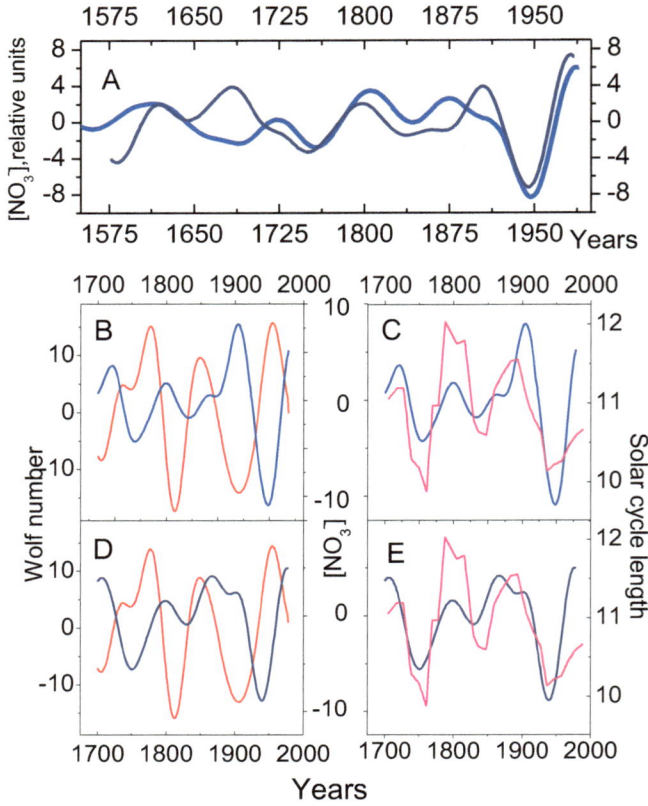

Fig. (5.14). Greenland nitrate record after Dreschhoff and Zeller [125] filtered in the 55-147-year scale band - blue line (**A**, **B**, **C**); Greenland nitrate record after Mayewski *et al.* [239] (φ=72.6° N, λ=38.5° W., altitude 3200 m a.s.l.) filtered in the 55-147-year scale band - dark blue line (**A**, **D**, **E**); Wolf number filtered in the 55-147-year scale band - red line (**B**, **D**); SCL smoothed by 5 points - pink line (**C**, **E**).

Ogurtsov *et al.* [264] have shown that century-type variations in Central Greenland NO_3^- records and corresponding periodicity in Wolf number (the cycle of Gleissberg) correlate negatively. For the dataset of Mayewski *et al.* [239] the coefficient of correlation is maximal (r_l= -0.71) when century-scale nitrate cycle leads the solar Gleissberg cycle by 17 years. The correlation is significant at 0.98 c.l (Fig. **5.14D**). The century-long periodicity of the nitrate record of Kansas University also correlates negatively with that of sunspot numbers (Fig. **5.14B**). The maximal value of the correlation coefficient (-0.58, 0.96 c.l.) is reached when nitrate leads solar activity by 12 years. Therefore, century-scale variations of NO_3^- abundance in polar ice show an apparent link with the solar Gleissbeg cycle, but with an unphysical phase shift (as well as in the case of instrumental temperatures). The coefficients of correlation between nitrate and smoothed solar cycle length, annually interpolated, are maximal (r_l = -0.53 to -0.59) for zero phase shifts.

Century-long periodicities in northern Fennoscandian (ϕ = 68-70° N) temperature and nitrate in Greenland ice are compared in Fig. (**5.15**) (reconstructions of Briffa *et al.* [309] and Lindholm and Eronen [306] were used).

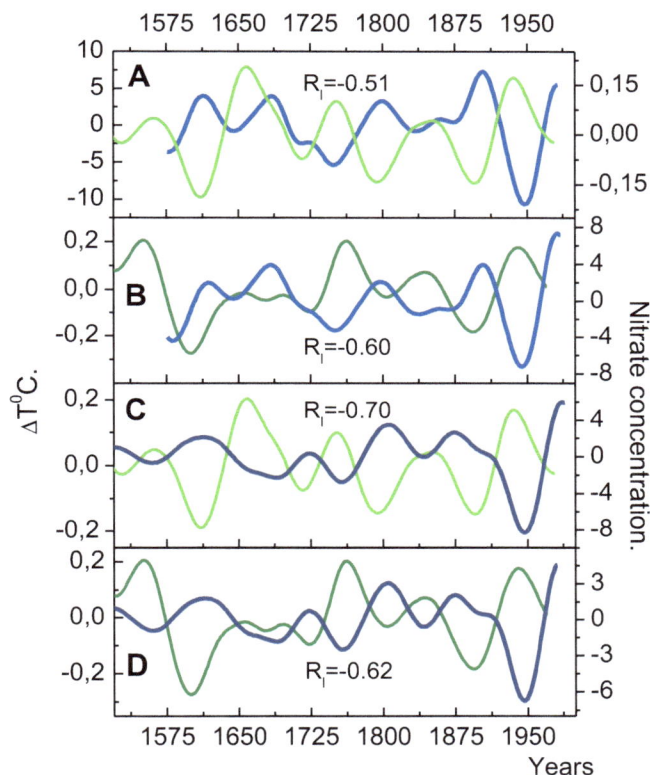

Fig. (5.15). Time series wavelet filtered in the 55-147-year scale band. Green line - northern Fennoscandian temperature proxy after Lindholm and Eronen [306]; dark green line - northern Fennoscandian temperature proxy after Briffa *et al.* [309]; blue line - Greenland nitrate record after Dreschhoff and Zeller [125]; dark blue line - Greenland nitrate record after Mayewski *et al.* [239].

A correlation between annually resolved temperature and NO_3^- records, wavelet-filtered in a Gleissberg scale band, exists during all of the last four to five centuries (Fig. **5.15**). Coefficients of correlation reach -0.51 to -0.70 (significant at 0.94-0.99 c.l.) with zero phase shifts.

Since nitrate in polar ice is linked to the stratosphere's ion production rate it is reasonable to suppose that:

(a)   the century-long cycle in stratospheric ionization leads the corresponding cycle in sunspot number;

(b)   the century-type periodicity of temperature of northern Fennoscandia is linked to the respective cycle in atmospheric ionization.

Ogurtsov [520] hypothesized that the century-type variation in stratospheric ion production rate could be a result of superposition of the century-long periodicities in GCR and SCR fluxes (Fig. **5.16**).

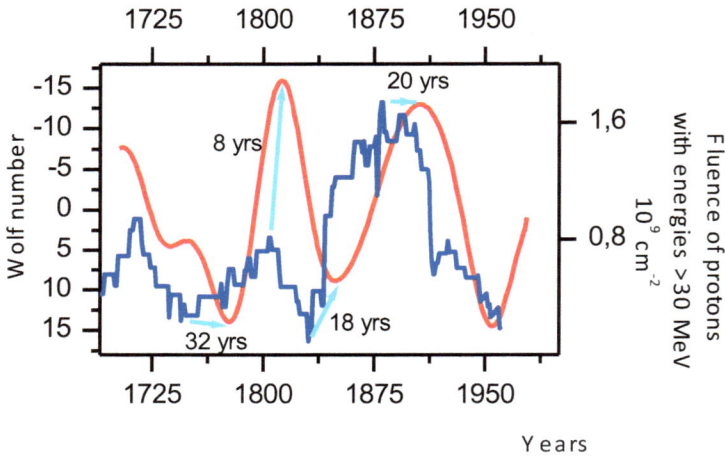

Fig. (5.16). The influence of powerful solar proton events reconstructed by McCracken *et al.* [257] averaged over 30 years - blue line; Wolf number filtered in a 55-147-year scale band (reverse scale) - red line.

Reconstructed SPE fluence has a century-type variation whose extremes considerably lead the corresponding extremes of a Gleissberg cycle in sunspot activity Fig. (**5.16**). Thus, century-long variation of stratospheric ion production, which leads the respective solar cycle in Wolf number, could be a result of the common action of: (a) the century-scale variation in GCR intensity, which develops in anti-phase with the corresponding sunspot cycle, and (b) the century-scale cycle in SCR flux, which also develops in anti-phase with the solar cycle but with the phase advancing. If ionization of the stratosphere indeed influences its transparency, a negative correlation between ionization (nitrate concentration) and surface irradiance (surface temperature) appears. Instrumental data give support for this assumption (Fig. **5.17**). Experimental evaluation of the aerosol optical depth of the atmosphere performed by Bryson and Goodman [521] by means of the data of 42 actinometrical stations situated within the 25-65° N belt.

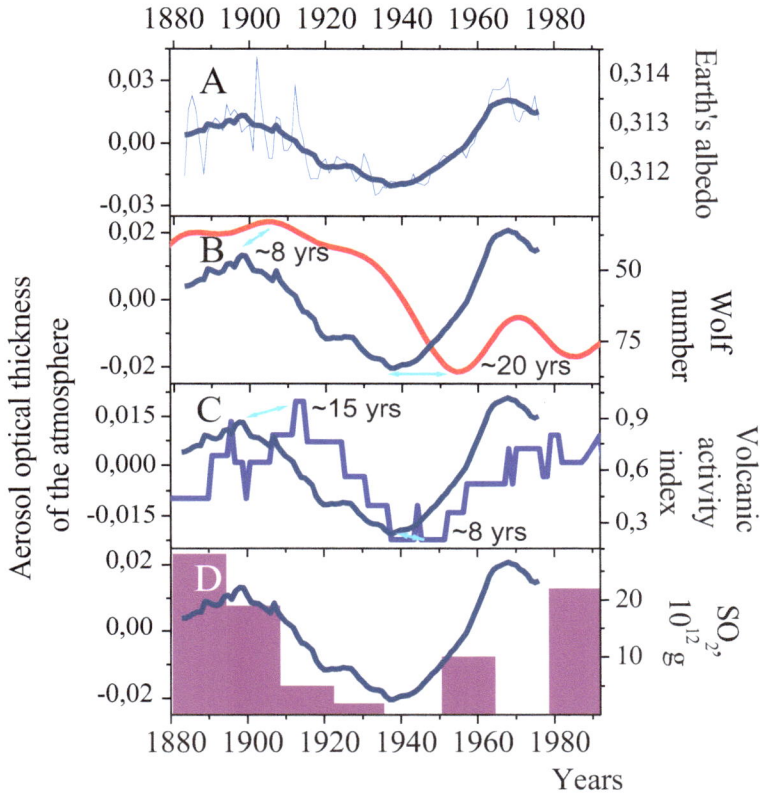

Fig. (5.17). Changes of atmosphere parameters. (**A**) aerosol optical thickness of a middle latitude ($\varphi = 25\text{-}65°$) stratosphere [521], annually interpolated. Thin blue curve - raw data; thick dark blue curve - data smoothed by 11 years; (**B**) red curve - Wolf number smoothed by the Fourier filter (frequencies above 0.04 years^{-1} are reduced); dark blue curve - aerosol optical thickness of the stratosphere, smoothed by 11 years; (**C**) violet curve - volcanic explosive index [312] smoothed by 25 years; (**D**) magenta columns - SO_2 loading to stratosphere [522].

Secular variations of atmospheric transparency in general are the result of changes in the optical thickness of the stratosphere (aerosols in troposphere are short-lived - see Table **5.6**). The century-long cyclicity in the stratospheric aerosol optical depth correlates negatively with the respective variation in sunspots and leads it by 20-25 years (Fig. **5.17B**). As noted above, volcanoes are the main source of sulfur dioxide in the stratosphere, from which sulphate aerosol is produced.

Multidecadal periodicity in volcanic activity, however, does not advance the corresponding variation in aerosol transparency but lags behind it (Fig. **5.17C**). The aerosol optical depth does not follow well volcanic sulfur dioxide loading into the stratosphere as evaluated by Kondratyev [522] (Fig. **5.17D**). This means that secular periodicity of the sulphate aerosol concentration in the stratosphere most likely is affected not only by volcanic SO_2 input but also by some supplementary factor. Ogurtsov [520] assumed that this factor might be a century-scale variation in stratospheric ion production. Ogurtsov [520] supposed that the positive time lag between century-type periodicities in surface temperature and sunspot number could be related not to the change in SCL but to the variation in stratospheric optical depth. That is, century-scale variability in terrestrial temperature could be at least partly a result of the respective change in the 'grey filter' efficiency. This again brings up a question about physical processes able to explain the relationship between aerosol concentration and ion production rate.

5.5. Ion-Mediated Nucleation of New Aerosols

An important source of new aerosols is nucleation from condensable trace vapours in the atmosphere. The most important of these is believed to be sulphuric acid. In the stratosphere only molecular clusters containing H_2SO_4 and H_2O molecules can form condensation nuclei (CN).

The mechanism of binary nucleation is a process of spontaneous formation of embryos of the liquid phase consisting of molecules H_2SO_4 and H_2O (Fig. **5.18**). Gas molecules participating in Brownian motion impact on each other. As a result temporary associations - clusters including several molecules - appear. Molecules are kept together in them due to cohesive forces.

However not all of these clusters are stable. The fate of each molecular complex is connected with the relation between the energy of surface tension and the latent heat released after cluster formation. The surface tension connected with coupling of molecules keeps their group in an association. The released heat of condensation raises the temperature of the cluster, increases the energy of its molecules, and promotes disintegration of this molecular complex. In simplified terms, the high-energy molecule which collides with a cluster can simply break it. The classical theory of nucleation [523] is based on the difference in the Gibbs free energy between: (a) the number of free molecules of the condensing

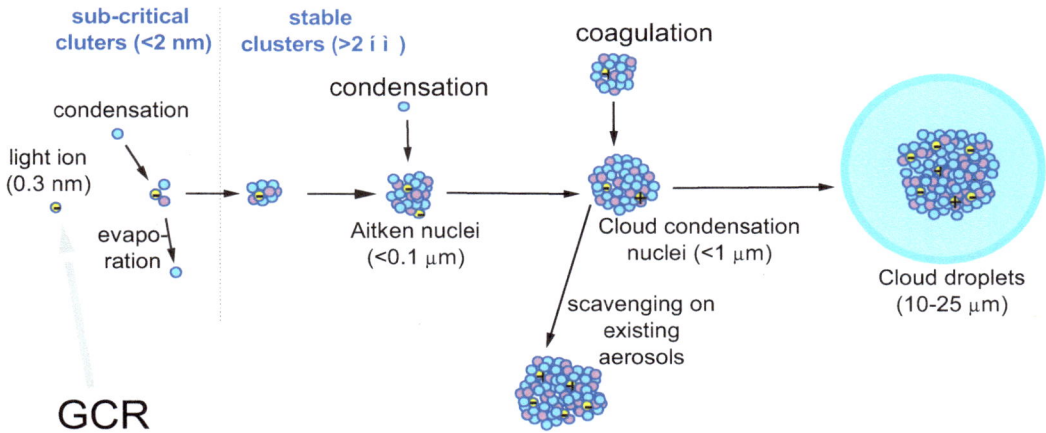

Fig. (5.18). Scheme of ion-mediated nucleation of new particles from condensable vapours in the atmosphere.

substance, and (b) the Gibbs free energy of these molecules when they are incorporated into a cluster. The development of occasionally occurring droplet embryos depends on the changes of free energy of Gibbs ΔG when free vapour molecules are transformed into molecules in the liquid state.

Gibbs ΔG consists of two parts: a negative part, proportional to r^3 (the heat of condensation), and a positive part, proportional to r^2 (the surface tension):

$$\Delta G = -kT \ln\left(\frac{P}{P_s}\right) \cdot n + 4\pi\sigma r^2, \tag{5.13}$$

where $n\ (\approx r^3)$ is the number of molecules in a cluster, P is the partial pressure of vapour in the atmosphere, P_s is the equilibrium vapour pressure over the surface of the liquid, σ is the surface tension, r is the radius of a cluster, and k is Boltzmann's constant. The molecular association is stable if $\Delta G < 0$. If $P < P_s$, the value ΔG is always positive and all clusters are unstable (as they tend to evaporate). In supersaturated vapour ($P > P_s$), at small radii the positive term dominates, but after the radius exceeds a critical value r_c the value ΔG becomes negative. The embryo, which is larger than r_c, is stable and can grow further. In the stratosphere, r_c of the complexes consisting of molecules of H_2O and H_2SO_4 is usually 1.5-2.0 nm. If the cluster is formed around an ion (ion nucleation) it appears charged. In that case the action of the Coulombic force of attraction reduces ΔG and promotes stabilization of the droplet. In the formula (5.13), the electrostatic term A_E is introduced:

$$\Delta G = -kT \ln\left(\frac{P}{P_s}\right) \cdot n + 4\pi\sigma r^2 + A_E,$$

$$A_E = \frac{q^2}{2} \cdot \left(\frac{1}{r_i} - \frac{1}{r}\right) \cdot \left(\frac{1}{\varepsilon_l} - \frac{1}{\varepsilon_v}\right),$$

(5.14)

where r_i is the radius of the ion, ε_l is the permittivity of liquid, ε_v is the permittivity of vapour, and q is the ion charge. Since $\varepsilon_l > \varepsilon_v$, the value of A_E is negative. The probability of formation of charged (more stable) clusters grows with atmospheric ionization. The formed stable molecular complexes of the supercritical size - aerosol embryos - can grow further by condensation and coagulation. The process of ion-induced nucleation can be described by a simplified formula [524]:

$$J_{sc} = \frac{Q}{\left(1 + \dfrac{\sqrt{\alpha Q}}{K \cdot [H_2 SO_4]}\right)^{N_c}},$$

(5.15)

where J_{sc} ($cm^{-3} \times s^{-1}$) is the nucleation rate (the rate of generation of super-critical particles with $r > r_c$), Q ($cm^{-3} \times s^{-1}$) is the ion production rate, α is the ion-recombination coefficient, K ($\cong 10^{-9}$ $cm^3 \times s^{-1}$) is the coefficient of the rate of H_2SO_4 molecules uptake by a charged cluster, $[H_2SO_4]$ is the sulphuric acid vapour concentration (cm^{-3}), and N_c is the minimal number of H_2SO_4 molecules in the critical cluster. The formula (5.15) describes the competition between: (a) growth of the ion cluster due to the association of H_2SO_4 molecules and (b) neutralization of the cluster ions due to collisions with other ions. The value $(K \cdot [H_2SO_4])^{-1}$ represents t_A, the time of association of the ion with the sulphuric acid molecule. If $[H_2SO_4] = 10^5 - 10^6$ cm^{-3} (typical values for the stratosphere), t_A is $10^2 - 10^3$ s. The characteristic time of ion-ion recombination is $t_r = (\alpha n)^{-1}$ ($n = n_+ = n_-$) and since in a steady state $Q = n^2 \cdot \alpha$ we obtain $t_r = \dfrac{1}{\sqrt{\alpha Q}}$. The coefficient of ion recombination α varies almost exponentially from 5.5×10^{-9} $cm^3 \times s^{-1}$ at 10 km to 6×10^{-8} $cm^3 \times s^{-1}$ at 30 km. It is evident that the rate of ion-mediated nucleation is highly dependent on the number N_c. The value of N_c, in turn, depends on stratospheric conditions: the temperature, relative humidity RH, and $[H_2SO_4]$. The background concentration of the sulphuric acid vapour is ca. 10^5 molecules per cubic centimetre at an altitude of 25 km [525] and can reach 10^6 cm^{-3} at 30-35 km. After volcanic eruptions, $[H_2SO_4]$ can exceed 10^7 cm^{-3} at 30 km [526]. Modern estimation of $[H_2O]$ concentration in the stratosphere gives values of 4-6 ppmv, which is equivalent to a relative humidity of ca. 1%. However, the dispersion of assessments is rather high, for example Khrgian [1] reported that a concentration of water vapour of 3.5×10^{-4} ($RH \cong 30\%$) was registered in the USA in 1959 at 26 km. Yu [527] has calculated values of N_c for different atmospheric parameters (Fig. **5.19**).

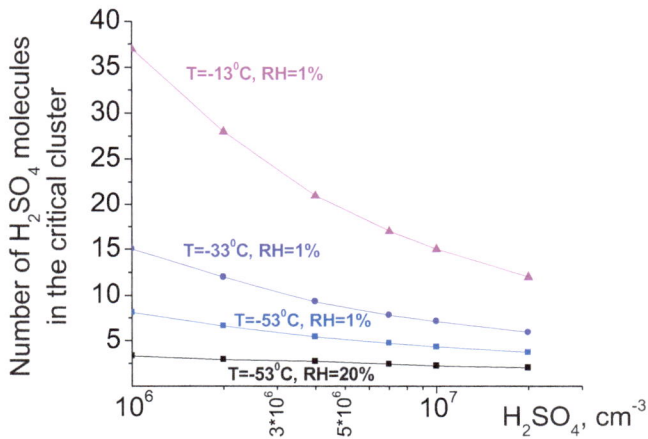

Fig. (5.19). The dependence of the number of H_2SO_4 molecules in the critical cluster on physical conditions in the atmosphere [527].

A more accurate formula by Kazil and Lovejoy [528] is convenient for theoretical studies of the process of ion-mediated nucleation in the stratosphere:

$$J_{sc} = \frac{Q}{\left(1+\dfrac{\sqrt{\alpha Q}}{K' \cdot \left[H_2SO_4\right]}\right)} \cdot \left(1+\frac{\sqrt{\alpha Q}}{2 \cdot K' \cdot \left[H_2SO_4\right]}\right) \cdot \frac{1}{\left(1+\dfrac{\sqrt{\alpha Q}}{K \cdot \left[H_2SO_4\right]}\right)^{Nc}}, \qquad (5.16)$$

where J_{sc} (cm$^{-3} \times$ s^{-1}) is the rate of generation of super-critical particles with $r > 3$ nm, and K and K' are the coefficients of H_2SO_4 molecules uptake by sub-critical ($r_c < 3$ nm) charged and neutral clusters. The formula (5.16) can be used to estimate the rate of ion-induced nucleation at heights of 25 and 35 km in different ion production rates and physical conditions (Fig. **5.20**).

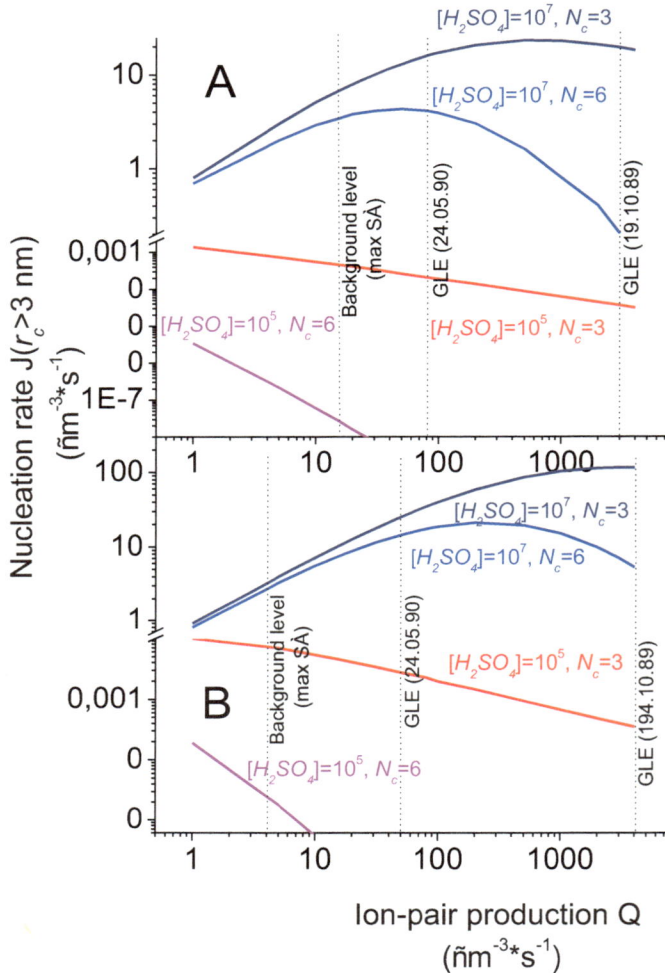

Fig. (5.20). Rate of generation of clusters with sizes of more than 3 nm at: (**A**): 25 km height ($\alpha = 4 \times 10^{-7}$ cm$^3 \times$ s^{-1}, $K = 10^{-9}$ cm$^3 \times$ s^{-1}, $K' = 6 \times 10^{-10}$ cm$^3 \times$ s^{-1}); and (**B**): 35 km height ($\alpha = 8 \times 10^{-8}$ cm$^3 \times$ c^{-1}, $K = 10^{-9}$ cm$^3 \times$ c^{-1}, $K' = 6 \times 10^{-10}$ cm$^3 \times$ c^{-1}).

The background rate of ion production does not exceed several tens of ion pairs per cubic centimetre per second even in the high-latitude stratosphere. However, during SPE this value can substantially increase in the polar ($\varphi > 60°$) stratosphere (Fig. **5.21**). Kasatkina and Shumilov [446] have calculated results of ionization (Table **5.9**). Data of ionization measurements in the atmosphere above Thule

(Greenland, $\varphi = 88°$ N) during the solar minimum [529, 530] were taken as background values.

Table 5.9. The rate of ion pair production Q in the high-latitude stratosphere.

Altitude (km)	Background Value Q Ion ($cm^{-3} \times s^{-1}$) (SA min)	Q During GLE (24 May 1990) ($cm^{-3} \times s^{-1}$)	Q During GLE (19 October 1989) ($cm^{-3} \times s^{-1}$)
25	15	$\cong 80$	$\cong 3 \times 10^3$
35	About 4	$\cong 50$	$\cong 4 \times 10^3$

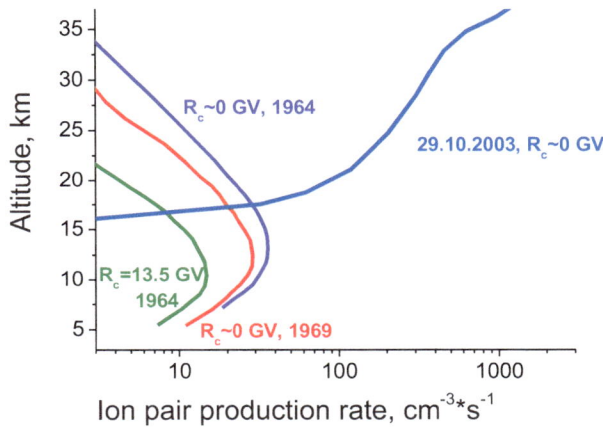

Fig. (5.21). Altitude dependence of the ion pair production rate. Violet line - above Thule ($\varphi = 88°$ N, $R_c \cong 0$ GV) during solar minimum; red line - above Thule during solar maximum; green line - at $R_c = 13.5$ GV during solar minimum; blue line - at geomagnetic pole after SPE of October 2003. Data adopted from [529, 531, 532].

The increase of stratospheric ionization after solar proton events (Fig. **5.20**) can actually result in a substantial rise (three times at 25 km, > 20 times at 35 km) of the rate of generation of new ultrafine particles if the following conditions are satisfied:

(a)   the concentration of sulphuric acid vapour is high enough ($\geq 10^7$ cm^{-3});

(b)   the number of sulfuric acid molecules in a critical cluster is low (≤ 4).

So a small N_c value of less than $[H_2SO_4] = 10^7$ cm^{-3} and $RH \cong 1\%$, typical for the stratosphere, requires temperatures of less than 220 K (-53 °C). It is evident that during the period of 16-23 February 1984, when Shumilov *et al.* [441] observed the generation of the aerosol layer above the Kola peninsula, the stratospheric conditions were favourable: (a) sulphuric acid concentration was elevated due to the El-Chichon eruption (April 1982, VEI = 6); (b) temperatures at 100 hPa level were < 200 K. Such low temperatures, however, are rather rare in the stratosphere during the warm season (April-September). One can note also that the connection between J and Q is non-linear. If ionization in the stratosphere is high ($Q >$ 200-300 cm$^{-3} \times$s^{-1}, strong proton flares), recombination can predominate, which results in a decrease of the ion nucleation rate with a rise of Q [446]. In spite of the quantitative character of the simplified estimations performed, it is possible to draw some conclusions:

(a)   An increase of ionization in the stratosphere caused by intrusions of high-energy SCR can increase the rate of generation of super-critical stable clusters consisting of H_2O and H_2SO_4 molecules substantially;

(b)   A link between stratospheric ionization and the rate of stable cluster generation J is likely to be non-linear; that is, weak SPE can sometimes increase J more effectively than strong ones;

(c)   A link between the ion production rate in the stratosphere and the rate of generation of clusters of super-critical size may depend drastically on physical conditions in the stratosphere.

Thus, the simplified estimations made above have shown that the ion-induced nucleation strongly depends on atmospheric weather conditions. Thereupon it is not unusual that aerosol enhancement in the stratosphere is observed after some SPEs but not after others. Estimations by Yu [533] showed that ion-mediated nucleation is also effective in the lower troposphere and thus can provide a positive correlation between cosmic intensity flux and low cloudiness.

Empirical evidence of ion-catalysed nucleation has been obtained:

1)   Hofmann and Rosen [534] performed a balloon study of the concentration of condensation nuclei at 25-30 km above Laramie (41° N) in 1979-1982. They described the *condensation nuclei events* - sudden increases of CN (mainly particles 0.01-0.1 μm in size), at least part of which could be associated with SCR bursts.

2)   Harrison and Aplin [535] measured the concentration of negative ions and condensation nuclei near the surface. They found that these values are positively correlated, which can be most easily explained in the frame of an ion-induced nucleation mechanism.

3)   Eichkorn *et al.* [536] detected large positive ion clusters with atomic mass numbers up to 2500 in the upper troposphere (9-10 km) above central Europe by means of an aircraft-based large ion mass spectrometer. Since there were not enough condensable gases in the atmosphere, attachment of aerosol particles to the ion clusters was considered the only way to explain the largest detected particles [536].

The most important experimental studies of the ion-mediated nucleation process were performed by a group of researchers from the Danish National Space Centre and by the CLOUD experiment at CERN.

4)   Danish researchers experimentally investigated the influence of ionization on the rate of formation of stable clusters, which can further transform into condensation nuclei. They used a 7 m^3 reaction chamber first [537] and then a much smaller 50 L chamber [538]. In both cases the chambers were filled with cleared humidified air containing O_3, SO_2, and water vapour at concentrations appropriate for the conditions in the Earth's atmosphere. The relative humidity of air in the chambers was 35-50%, the pressure was kept at 1 mbar above atmospheric pressure, and the temperature was 20-23 °C. Additional ionization was created by means of a gamma radiation source [537]. In the 7 m^3 chamber, the generation of new aerosol particles was found to be proportional to the negative ion density. For example, at $[H_2SO_4] = 2 \times 10^8$ cm^{-3}, the concentration of particles with a size > 3 nm increased three times when the concentration of ions in the chamber increased from 1800 up to more than 5000 cm^{-3}. Similar results were obtained in a series of experiments with a 50 L chamber at sulphuric acid concentrations of 0.5-7 $\times 10^8$ cm^{-3}.

5)   The CLOUD experiment at CERN [539] was performed using a 26.1 m^3 chamber. Unlike the Danish experiments the mixture pumped into the chamber contained not only cleared humidified air, sulphur dioxide, and ozone (O_3) but also ammonia vapour. The mixture was very clear; that is, it had much lower concentrations of contaminants than all previous experiments. The temperature during the experiment varied in the range of -90 to +27° C. A beam of secondary pions produced by a proton synchrotron

was used as an ionization source. The CLOUD chamber conditions were very close to those in the atmosphere, and thus a comparison of the nucleation rate detected in the experiment with the rate actually measured in the atmosphere was possible. The experiment confirmed the influence of ionization on the nucleation rate, especially under conditions characteristic for the middle troposphere ($T = -25$ °C, $RH = 38\%$, $[H_2SO_4] = 0.6\text{-}2 \times 10^7$ cm^{-3}). However, the results obtained under the conditions of the lower atmosphere were rather unexpected. The rate of nucleation observed in the chamber was only $10^{-1}\text{-}10^{-3}$ of the rate observed in the lower atmosphere. Thus the most likely nucleating vapours H_2SO_4 and NH_3 cannot account for the nucleation that is observed in the lower atmosphere. This means that nucleation processes in the atmosphere still remain unclear and need further investigation. Most likely some other gaseous impurities, besides sulphuric acid and ammonia vapour, take part in the process of generation of new ultra-fine particles in the atmosphere.

Thus, the rise in stratospheric ionization can considerably increase the rate of formation of aerosol embryos by means of the ion nucleation mechanism. The newly nucleated aerosol droplets can grow further by coagulation and condensation. It is important to evaluate the time required for a superfine particle to reach the sub-micron size since such particles (a) scatter the sunlight most effectively (their size is close to the wavelengths of visible light), and (b) can be detected by lidar. Aerosol coagulation is a process in which small particles collide due to their random motions, join together, and form larger particles. According to the majority of theoretical estimations (*e.g.* Zuev and Kabanov [540]), the coagulation process has a very poor influence on the growth of particles with radii of more than 0.1 μm in the atmosphere. Hamill and Toon [541] have estimated that under typical stratospheric conditions, even after two years of coagulation, the number of particles with a size > 0.1 μm will change very slightly. Condensation is a more effective mechanism for aerosol particle growth in the stratosphere [541]. The process of growth of the formed embryos due to condensation of the sulphuric acid vapour can be described by the equation of Hofmann and Rosen [526]:

$$\frac{dr(t)}{dt} = \frac{m \cdot D \cdot (n(t) - n(t_0))}{r(t) \cdot W \cdot \rho \cdot (1 + \lambda \, K_n)}, \tag{5.17}$$

where r is a radius of aerosol particle; $m = 1.7 \times 10^{-22}$ g is a mass of an H_2SO_4 molecule, W is an acid weight fraction; ρ is the aerosol (H_2SO_4 solution) density;

$Kn = (l_{eff}/r)^{-1}$ is a Knudsen number, $l_{eff} = \dfrac{1}{\pi \, N_{air} \, d^2 \, \sqrt{M_{air}/(M_{air} + M_{H_2SO_4})}}$ - is an

effective free path for the sulfuric acid molecules in the air, N_{air} is the density number of air molecules, $M_{air}=29.0$ g $\times$mol^{-1} is the molecular weight of air,

$M_{H2SO4}=98.1$ g $\times$mol^{-1} is the molecular weight of an H_2SO_4 molecule; $d=4.3\times10^{-8}$ cm is the mean collision diameter of an H_2SO_4 molecule with air, $\lambda = \dfrac{4/3 + (\sqrt{2}Kn) - 1}{1 + 1/Kn} + \dfrac{4 \cdot (1-\alpha)}{3\alpha}$ is a correction factor which depends on Kn and

sticking coefficient, α is a sticking coefficient, $D = \dfrac{1}{3} \cdot \left(\dfrac{8kT}{\pi m} \right)^{1/2} \cdot l_{eff}$ is a

diffusion coefficient of H_2SO_4 molecules [542], $n(t)$ is a concentration of $[H_2SO_4]$ vapor, t_0 is a moment when the condensation starts.

The concentration of H_2SO_4 in stratospheric air should depend on time. Actually, the more sulphuric acid vapour has settled on the existing droplets, the fewer H_2SO_4 molecules remain in the atmosphere. Hofmann and Rosen [526] described this dependence using the simple formula:

$$n(t) = n(t_0) \cdot \exp\left(-\frac{t - t_0}{\tau} \right). \tag{5.18}$$

As the basic source of stratospheric H_2SO_4 vapour is sulphur dioxide it is reasonable to use τ as the lifetime of SO_2 in the stratosphere ($\tau = $ 30-100 days). However, if there is a permanent and copious input of SO_2 to the stratosphere, the concentration of the sulphuric acid vapour can remain constant ($\tau = ¥$). Numerical solution of equation (5.17) with the properties of stratospheric aerosol $W = 0.8$, $\rho = 1.8$ g $\times$ cm^{-3}, $T = 220$ °K, $\alpha = 1$ [526], $t_0 = 0$, and the following initial condition

$$\left. \frac{dr(t)}{dt} \right|_{t=0} = \frac{1}{4} \cdot \left(\frac{8kTm}{\pi} \right)^{1/2 \cdot} \frac{n(0)}{W\rho}, \tag{5.19}$$

results in the outcome (Fig. **5.22**) that, even if the concentration of sulphuric acid is high ($[H_2SO_4] = 10^7$ cm^{-3}) and constant in time ($\tau = ¥$), a cluster of a few nanometres in size reaches a diameter of 0.3 μm after 2.5 months and 0.7 μm after half a year of condensation. If conditions are not so favourable for condensation ($[H_2SO_4] = 10^6$ cm^{-3}) this amount of time will increase many times. If it is further taken into account that the decrease of [H_2SO_4] due to condensation ($\tau = 100$ days) aerosol particles do not in principle reach the value of 0.7 μm (the size limit of a lidar registration in the experiment by Shumilov). This was observed by Hofmann and Rosen [526], who measured the sizes of aerosol particles over Laramie after the El-Chichon eruption. During ca. 200 days after the eruption the mean radius of particles grew from 0.01 to 0.17 μm at 17-20 km and to 0.27 μm at 20-25 rm. Then further growth ceased. These results do not agree well with the fast (a few days) response of the stratospheric sub-micron aerosol to solar proton events registered in several experiments [441, 449, 451]. It should also be noted that fast precipitation of aerosol layers with speeds of 1-2 km/day [441] and 5 km/day [442] was observed. This means that the speed of sedimentation of particles present in these layers reaches 1-5 cm × s^{-1}.

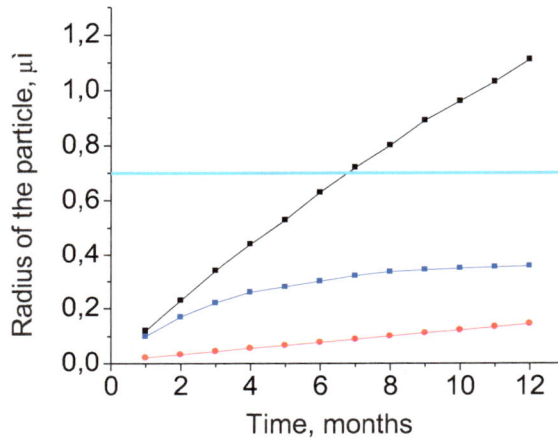

Fig. (5.22). Theoretical time behaviour of aerosol particles' growth in the lower stratosphere ($W = 0.8$, $\rho = 1.8$ g × cm^{-3}, $T = 220$ °K, $\alpha = 1$). Black line - [H_2SO_4] = 10^7 cm^{-3}, $\tau = \infty$. Red line - [H_2SO_4] = 10^6 cm^{-3}, $\tau = \infty$. Blue line - [H_2SO_4] = 10^7 cm^{-3}, $\tau = 100$ days.

The velocity of sedimentation of a particle could be described by the formula:

$$v = \frac{2 \cdot g \cdot (\rho - \rho_{air}) \cdot r^2}{9v}, \qquad (5.20)$$

where v is expressed in cm × s^{-1}, $g = 980$ cm × s^{-2} is the gravitational constant, ρ is the density of a particle, ρ_{air} is the density of air, r is the radius of a particle in centimetres, and v is the dynamic viscosity of air in g × s^{-1} × cm^{-2}.

Using formula (5.20) we find that sedimentation velocities of 1 and 5 cm $\times$ s^{-1} give particles with radii of r = 6 and 14 μm, respectively. Thus the lidar experiments performed by Shumilov *et al.*, [441] and Marichev *et al.* [442] provide evidence that large ($\cong$ 10 μm) aerosol particles can originate in the stratosphere during only a few days after sharp increases of ionization. Thus fast generation of such large particles cannot be explained by ion-catalysed nucleation.

5.6. Other Possible Mechanisms of a Cosmic Ray-Aerosol Connection

Yu [543] proposed a mechanism of quick formation of 10 μm particles of polar stratospheric clouds by means of electro-freezing - a *cosmic-ray-induced freezing* (CRIF). Polar stratospheric (or nacreous) clouds (PSCs) are another type of stratospheric aerosol (Fig. **5.23**).

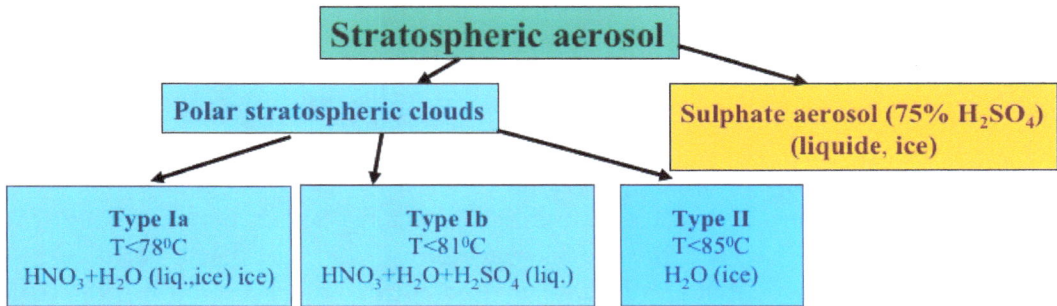

Fig. (5.23). The basic types of stratospheric aerosol.

These two types of PSCs both form under extremely low temperatures that can occur only in the winter in the Arctic and Antarctic stratosphere. Type II PSCs consist of water/ice particles with a size of 2-25 μm that form when the temperature falls below -85 °C. These PSCs occur mainly in the Antarctic since in the Arctic stratosphere such low temperatures occur rather seldom. Type I PSCs can be formed in warmer conditions and thus they emerge in the Arctic as well. It is believed that type Ia PSCs are composed of nitric acid trihydrates $HNO_3 \times 3H_2O$ (NAT), while type Ib PCSs are composed of supercooled liquid droplets of the ternary solution HNO_3-H_2SO_4-H_2O. According to Tolbert [544] this ternary solution consists of 40% HNO_3, 56% H_2O, and 4% H_2SO_4 at -83 °C. The frost points for Type I PSCs are -78 °C (Ia) and -81 °C (Ib). Particles of type Ia PSCs typically have a diameter of 2-5 μm while type Ib PSCs consists mainly of sub-micron particles. The optical depth of PSCs can reach 0.1 [545]. Nacreous clouds can cover up to 30% of the territory north of 60° N [546] and influence atmospheric transparency over vast areas. For example, sharp increases of optical depth (from $1\text{-}2 \times 10^{-3}$ to $6\text{-}8 \times 10^{-3}$) were observed over the Antarctic (64-80° S) during a cold period. Similar effects - rises of optical depth from 6×10^{-3} to 10^{-2} during winter - were detected in the Arctic (64-80° N) by Mohnen *et al.* [547]. The authors considered PSC generation as a cause of the observed transparency changes. Yu [543] developed a mechanism of cosmic-ray-induced freezing of super-cooled and thermodynamically unstable $HNO_3\text{-}H_2O\text{-}H_2SO_4$ droplets. The CRIF mechanism involved a possible reorientation of polar molecules in the aerosol solution and they transfer into crystalline structure induced by strong electrical fields of moving secondary ions generated by passing cosmic ray particles. Simulations by Yu [543] showed that CRIF can produce substantial enhancement of NAT particles over 15-22 km height during 2-6 days after strong SPE. The diameter of the particles can reach >10 μm at 16-18 km altitude. Yu [543] assumed that the substantial increase in the aerosol backscatter ratio registered by Shumilov *et al.* [441] was a result of PSC generation by means of the CRIF mechanism. It should be noted that the temperature at 100 hPa (about 20 km) above the Kola Peninsula was actually low on 16-21 February 1984. It did not exceed -75 °C during the entire time interval, and it dropped to -84° C on 16 February [441]. Cold conditions favourable for nacreous cloud generation were also quite probable in the Antarctic stratosphere during September-October 1989 when Mironova and Usoskin [451] observed the fast aerosol response to SPEs. However it is unrealistic that any PSC formation above Tomsk in March 1988 and 1989 or above the Antarctica in January 2005 appeared. Other processes like electro-coalescence and electro-scavenging may play a role here [480]. Nevertheless the physical mechanism linking space weather phenomena and aerosol in the stratosphere is still not fully understood. That is why investigations of solar-terrestrial relations continue.

Avakyan and Voronin [548] have suggested a new *radio-optical mechanism* of solar effect on weather and climate. They consider the effect of solar flares and geomagnetic storms on the state of the lower atmosphere as a three-step process:

(a) The fluxes of electromagnetic and corpuscular radiation caused by solar flares and precipitations from the radiation belts and magnetosphere during geomagnetic storms transform into a flux of microwaves (0.1-100 cm) in the ionosphere through the excitation of Rydberg states.

(b) The generated microwave radiation freely penetrates to the Earth's surface and influences the rate of formation and destruction of water cluster ions.

Waves with lengths of 2-5 cm promote the association of water vapour in clusters while waves with lengths of 5-10 cm facilitate disintegration of the cluster.

(c)   The water clusters, controlled by ionospheric microwave radiation, in turn affect the formation of cloud and aerosol layers, influencing the surface radiation flux.

The radio-optical mechanism looks plausible but needs more experimental support and some numerical evaluation.

Koudriavtsev and Jungner [549] proposed a possible mechanism by which cosmic rays affect the formation of neutral water droplets and ice crystals in the Earth's atmosphere. This mechanism is based on changes in atmospheric transparency and vertical temperature distribution. They showed that a change in the atmospheric optical thickness in visible and IR range, which can take place when cosmic-ray particles enter the atmosphere, results in a change in the vertical distribution of temperature, appreciably affecting the growth of water droplets, concentration of active CN, and formation of ice particles.

5.7. Summary and Remaining Problems

A lot of empirical evidence of a relationship between the solar-modulated cosmic ray flux and surface irradiance (the CR-SR$\uparrow\downarrow$ effect) has been obtained over the last decades. This relationship manifests itself both over short-term (days) and long-term (inter-annual to decadal) time scales. The effect is not robust - it can change with time (*e.g.* GCR-low cloud correlation) and has geographical and altitudinal variability. Moreover, some of the results contradict each other. Palaeodata, however, give support for the existence of the CR-SR$\uparrow\downarrow$ connection at multi-decadal to centennial time scales at least over some regions. The physics of this linkage is currently under intensive investigation, and the most probable mechanisms can be suggested (Fig. **5.24**).

Ion-mediated formation of newly formed particles looks like a plausible explanation for a long-term connection between cosmic rays and (a) aerosol in stratosphere and (b) clouds in troposphere, which has been reported in many works. But ion-catalysed nucleation cannot explain fast (within several days) enhancement of sub-micron stratospheric aerosol after the strong space weather events detected by Shumilov *et al.* [441], Marichev *et al.* [442], Mironova and

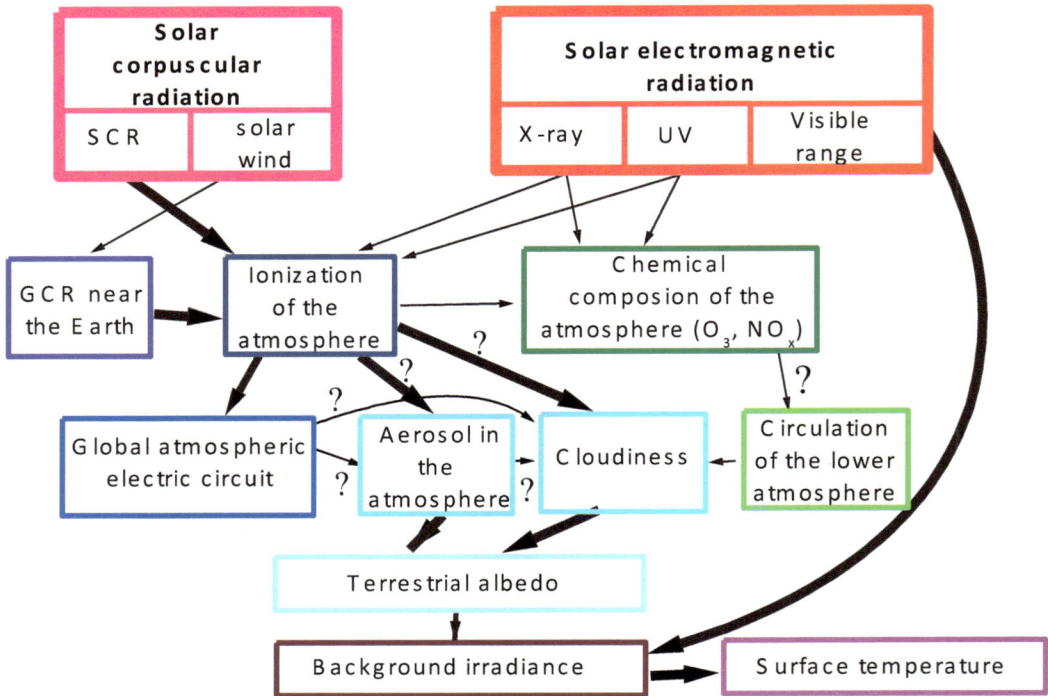

Fig. (5.24). Scheme of the possible mechanisms of the solar influence on weather and climate.

Usoskin [451], Veretenenko *et al.* [449], and. In reality the super-fine particles originating in the stratosphere can reach sub-micron size only after many months of condensation and coagulation. Quick formation of polar stratospheric clouds due to CRIF could be responsible for the increase in aerosol backscattering ratio observed by Shumilov *et al.* [441] shortly after the SPE of 16 February 1984. However, PSCs can appear only under very low temperature (T < -78° C) which probably occurred above Verkhnetulomsk (φ =68° N) in February 1984 but are very unlikely to have occurred above Tomsk (φ = 56° N) in March 1988, 1989 [442] or above the Antarctic during the warm season (January 2005, [451]). It should be noted that both ion-mediated nucleation and CRIF depend critically on the physical conditions of the atmosphere (concentration of sulphuric and nitric acids, humidity, temperature), which are determined by natural terrestrial processes (weather, volcanism).

Serious evidence of a connection between solar activity and atmospheric circulation was obtained recently. It was revealed that atmospheric pressure and circulation react to changes in cosmic ray intensity and solar activity over both short-term [468,469]) and long-term [472,477] time scales. Here as well as in the previous case the reaction is rather intricate: it can change sign with time and has a complex latitude-regional pattern. Veretenenko and Ogurtsov [472] have shown that the sign reversals of SA/GCR effects on atmospheric pressure could be a result of transformations of the large-scale atmospheric circulation. Changes of macro-circulation are in turn determined by inherent climatic variability. The physical mechanism providing a connection between atmospheric circulation and the Sun's activity is probably connected with perturbations in the radiative-thermal balance of the atmosphere due to: (a) variations of cloudiness and aerosol content associated with variations of ion-production rate and changes of atmospheric electricity and (b) variations in stratospheric ozone chemistry caused by UV solar radiation (Fig. **5.24**). Another way to provide a connection between solar luminosity and climate is variation in UV solar radiation, which effectively influences the ozone layer. Ultra-violet radiation has 11-year variation in reaches with the amplitude of the several percent (ca. 6% at 20 nm). Ozone effectively influences the radiative balance of the stratosphere with indirect effects on the circulation pattern of the troposphere pattern. The potential response of the climatic system to changes of UV solar radiation was analysed by Haigh [550], Shindell *et al*. [551], Gray *et al*. [465].

Finally, we can conclude that in spite of a number of plausible hypotheses, some of which has both theoretical and experimental support, the physics of the solar-climatic connection is still not well understood. The statistics of the obtained results remain insufficient, and this often prevents the drawing of unequivocal conclusions. For example, some authors revealed a strong response of the atmosphere to Forbush decreases of GCR intensity [452] while other authors did not find any reaction [453]. We can note that in this framework of modern consideration it is likely that solar activity does not affect the atmosphere directly but that its influence is mediated by a few internal terrestrial phenomena - volcanic activity, meteorology, and so on. The effect of solar activity on the atmosphere is probably non-linear. For example, ionization in the stratosphere induced by proton fluxes from SPEs can effectively facilitate new particle nucleation if the ion production rate is below some critical value, but if it exceeds the critical level the nucleation rate is reduced. Such a complex solar-atmosphere link might be the main cause of instability and regional-scale variability of solar-climatic correlations. Moreover, the solar-climatic connection might be mediated by natural factors of terrestrial origin (see Fig. **5.25**). *e.g.* both ion-induced nucleation and generation of nacreous clouds depend significantly on physical conditions in the atmosphere. Solar activity can influence the stratospheric aerosol and hence the weather on Earth only when temperature is low and the concentration of sulphuric acid is high. These values, however, are controlled by natural meteorologic processes and volcanic activity.

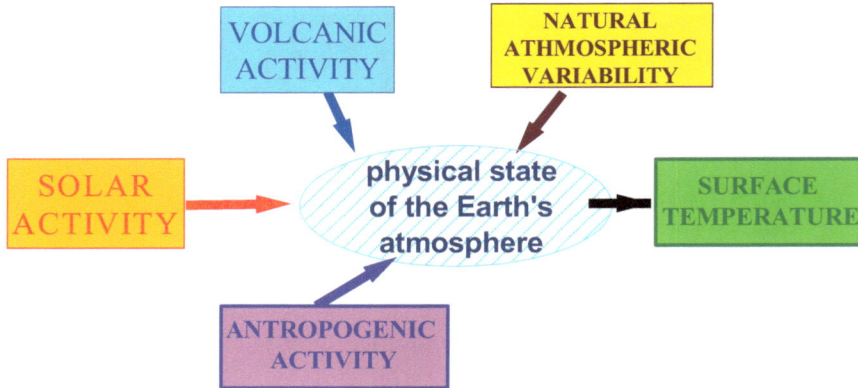

Fig. (5.25). A scheme of possible influences of different processes on the Earth's climate.

That is why thorough and continuous further monitoring of a variety of solar, space weather, and climatic phenomena using background, air-borne, and space-borne instruments is of paramount importance for understanding the solar-climate relationship. This is of particular importance now, because the long-term change of the Sun's activity is a possible contributor to global warming throughout the multiple effects of climate change on biological resources.

CHAPTER 6

Global Warming - Facts and Problems

Abstract: Possible factors contributing to global warming during the last 100 years - the greenhouse effect, natural climatic variability, variations of solar activity, and cosmic ray fluxes as well as anthropogenic changes in land-cover are described. It is shown that currently it is very difficult to estimate even the relative role of each of these factors in global warming due to our still inadequate knowledge of them as well as discrepancies between different key data sets. It is shown that despite some successes and tremendous efforts in the fields of both climatology and paleoclimatology our knowledge of the Earth's past climate is still lacking.

Keywords: Climatology, global warming, paleoclimatology.

Global warming (GW) has become the most important problem of climatology since the end of 80s. In the later part of the last century it was finally realized that the global climate is warmer than during any period over the past one hundred years. The theory of greenhouse effect and climatic modelling started to develop actively, and the Intergovernmental Panel on Climate Change (IPCC) was founded by the United Nations Environment Programme and the World Meteorological Organization. The aims of this organization are to compile the scientific knowledge of the hazards of human-induced variability of climate change and predict the impact of the industrial activity according to updated climate models. As noted in the Introduction, general awareness of the dangers of global warming has already transformed the subject to a topic of both global and regional political debates. A more complete understanding of GW and its mechanisms can provide us with important information about possible climatic variations in future decades and, therefore, is of crucial importance the well-being of mankind. For example, if the climate was warmer at some point in the last millennium than now (*e.g.* the Medieval Warm Period) then the lack of serious disasters during that period makes it reasonable to suggest that similar future warming would not be too devastating to humankind.

As the concentration of carbonic dioxide in the atmosphere has gradually increased since the 19th century due to human activity, it is reasonable to assume that mankind is responsible for GW. This hypothesis has got a great popularity among the experts in climatology. Recently it has been assertively declared that:

Maxim Ogurtsov, Risto Jalkanen, Markus Lindholm and Svetlana Veretenenko

a)   global warming is caused predominantly by human influence [150, 339, 401] and

b)   mean global temperature over the second half of the 20th century were probably (> 0.90 c.l.) higher than during any other 50-year period in the last 500 years and probably (> 0.66 c.l.) the highest in the past 1300 years.

These statements have in fact become the prevailing paradigm (world view or dogma) among the majority of climatologists and they form a basis for a number of political decision-making processes in industrialized societies. However, at the same time some loud sceptical arguments have become popular through various channels on the Internet and challenge the mainstream science [552]. Therefore, it is important to examine the specified assumptions using a variety of direct and proxy data and estimate the possible contribution of these other factors, including solar and cosmophysical ones.

6.1. Global Warming in the Context of Direct and Proxy Data

The IPCC consortium reported that the global mean surface temperatures measured by weather station thermometers rose by 0.854°C ± 0.20°C during the period 1880-2012 [401] (Fig. **6.1A**). At present it is a widely held a point of view that the global the temperature increase is: (a) mainly a result of human inputs of greenhouse gases (CO_2, CH_4, N_2O, halocarbons) and (b) is highly unusual and unprecedented in a historical context (*e.g.* [339]).

However, it is also generally acknowledged that the temperature records measured by thermometers are not representative enough to support definite conclusions. Even in the last 30 years the meteorological station network has covered less than 90% of the land surface, *i.e.* less than 25% of the total Earth's surface (Fig. **6.1B**). Furthermore, the scarcity of spatial coverage of the temperature records–rises going back in time. As a result the uncertainty of the estimation of global annual temperature, measured instrumentally, increases from about 0.1°C at the end of the 20th century to ca 0.15°C at the end of the 19th century to 0.18°C in AD 1850 [553].

The measurements of surface temperature by means of the satellite-born instruments started at the end of the 1970s. These records have a much denser spatial coverage particularly over the oceans. Satellites pass over most points on the Earth (both land and ocean) twice per day. Microwave sounding units (MSU)

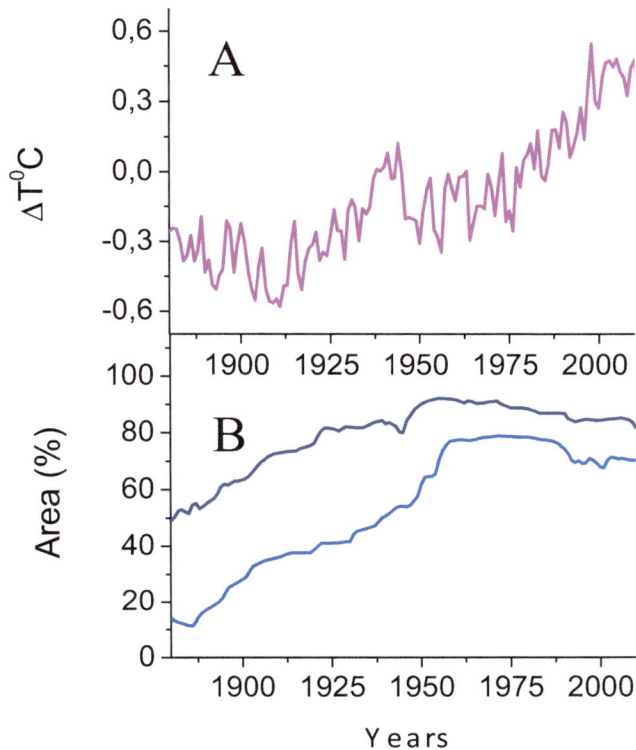

Fig. (6.1). Global temperature. (**A**) observed changes in a global average annual temperature measured by weather stations (http://www.cru.uea.ac.uk/cru/info/warming/; (**B**) - the percentage of hemispheric area situated within 1200 km of a reporting weather station. Dark blue line - Northern Hemisphere; blue line - Southern Hemisphere. Data were electronically scanned from http://data.giss.nasa.gov/gistemp/station_data/#form and digitized.

- space-borne microwave sounders – detect microwave thermal emissions (radiances) of ^{18}O molecules from a number of emission lines around 60 GHz. By taking measurements at different frequencies near 60 GHz ($\cong 1$ cm) temperature of different atmospheric layers can be estimated. Atmospheric temperature can then be calculated by means of various mathematical procedures. Two teams of researchers - the Remote Sensing System (RSS) group and the University of Alabama (UAH) group - have analysed the data obtained by the NASA satellite series TIROS (Television Infrared Observing Satellites). The RSS and UAH groups have obtained two versions of temperature changes in the lower troposphere, *i.e.* at altitudes less than 8 km (maximum of detector sensitivity was around 2-3 km), since the end of 1978 (Fig. **6.2**).

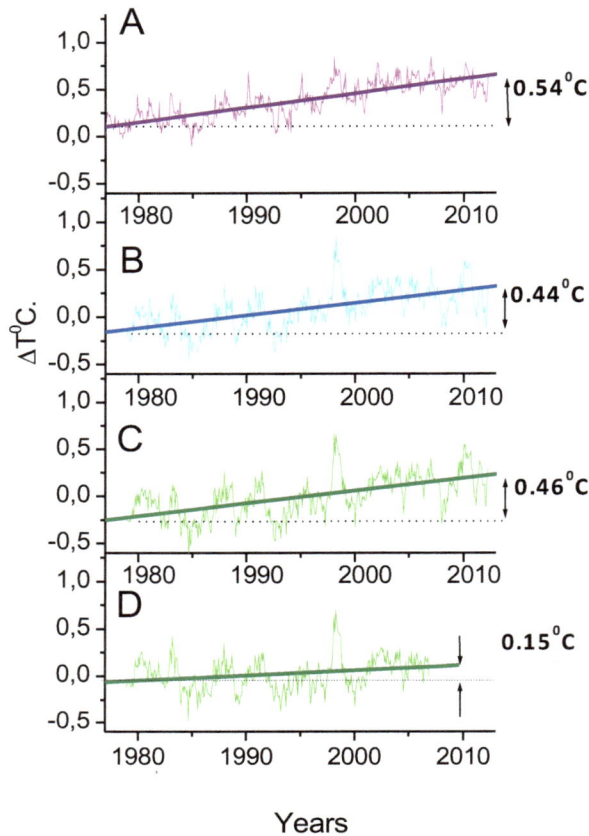

Fig. (6.2). The Remote Sensing System (RSS) and two University of Alabama (UAH) satellite temperature records (versions 2007 and 2012) together with surface thermometric data. (**A**) observed changes in global average temperature measured by weather stations (ftp://ftp.ncdc.noaa.gov/pub/data/anomalies/monthly.land_ocean.90S.90N.df_1901-2000mean.dat); (**B**) RSS satellite-based global lower troposphere temperature anomaly (http://www.remss.com/data/msu/); (**C**) UAH global lower troposphere temperature anomaly measured by satellite-based instruments (2012 version, http://vortex.nsstc.uah.edu/public/msu/); D-UAH satellite-based global lower troposphere temperature anomaly (2007 version, http://www.ncdc.noaa.gov/oa/climate/research/msu.html).

Combining readings from different satellites is a demanding task. Space-born measurements, however, have a few important limitations [554]: (a) offsets and drifts in satellite calibration, and (b) orbital decay and drift and corresponding long-term variations in the time of day of the observation at a particular location. That is why it is difficult to estimate correctly long-term trends in the space-born - data. The first versions (before 2005) of UAH record (Fig. **6.2D**) have a weak trend of 0.03-0.05°C/decade [555]. In 2005, the trend value changed to 0.12°C/decade [556].

The updated version of UAH series (Fig. **6.2C**) has a trend value of about 0.14°C/decade, which is similar to that in the RSS series (Fig. **6.2B**). Despite some past differences, both UAH and RSS series show that the temperature rise in the low troposphere during the last 30 years is likely higher than at the surface contrary to the forecasts of the greenhouse warming theory. Indeed, the general circulation models (GCMs) forecast that the low layers of troposphere will warm more rapidly than the surface due to the greenhouse effect [339]. According to GSM estimations this effect is most expressed in the low-latitude troposphere and therefore this part of the atmosphere gives the best perspective for disclosure the greenhouse fingerprint. The IPCC GCMs predict a greenhouse warming of the tropical troposphere, which increases with height and reaches its maximum at ca. 10 km (see Fig. **9.1** of [557]). Comparison of the model-predicted temperatures and - temperatures derived from satellite observations has produced rather controversial results that generated a decades-long discussions which continues even today. For example, Douglass *et al.* [558], who studied tropospheric temperature trends using 22 GCMs, have shown that the trends in model and observed temperature differ significantly in the largest part of the tropical troposphere. However, Santer *et al.* [559], who studied the updated and corrected temperature data sets, claimed, that the trends are within the confidence intervals of the models. Fu *et al.* [560], in turn, showed that GCMs significantly exaggerate the value of trends in the difference between the tropical upper troposphere and lower-middle troposphere temperatures in 1979-2010.

Ogurtsov *et al.* [365] analysed the nine temperature paleorecords covering the last 600-1000 years and concluded that the 20th century was actually the warmest over the last 500 years. A statistical analysis of eight paleoclimatic reconstructions, corrected for ARS effect, showed that in the extratropical part of the Northern Hemisphere the time interval 1988-2008 was the warmest 20 years during the last millennium with a probability of more than 0.70. On the other hand, McIntyre and McKitrick [393] have shown that at the beginning of the 15th century the global temperature was higher. Despite some differences between results obtained by means of different temperature reconstructions they provide serious evidence that at least the last 20 years were a period of unusually high temperature and hence a period of an unusual state of the climatic system. That, in turn, testifies that at the end of the 20th century the climatic system was disturbed by some additional forcing factor of a global scale. Thus, analyses and reanalyses of the available long-scale temperature proxies shows that the global temperature averaged over the last 100 years was actually higher than the global temperature averaged over the last 1000 years.

6.2. Modern Climate Modelling - Advantages and Limitations

Climate models describe the Earth's climate system by means of sets of mathematical equations and numerical methods. The models reproduce interactions of the atmosphere, ocean, cryosphere and land surface. Climate modelling has a number of aims from studies of the dynamics and evolution of the climate system to the forecasting future climate. Climate models vary in complexity from the simplest, energy-balance analytic models to state-of-the-art general circulation models (GCMs), reproducing the physics, chemistry and biology of many of the parts of the global climatic system.

In the frame of the energy balance approach, changes in the globally averaged climate system are evaluated from an analysis of the variations in the Earth's heat budget. The rationale behind these models was presented by Budyko [561] and Sellers [562]. Energy-balance model generates global mean values for the variables in question.

The more complicated radiative-convective models allow study the vertical transfer of energy through the Earth's atmosphere. This approach enable analysis of the relative the role of greenhouse gases, convection, water vapour, clouds and the stratosphere. Manabe and Wetherland [563] presented the first radiative-convective model.

GCMs return three-dimensional models, which have boundary conditions at the spreading surface. They are aimed at simulation incoming and outgoing radiation, effect of greenhouse gases, time/spatial variation of the wind field, generation of clouds and transfer of water vapour, variations of sea-ice cover, atmosphere-ocean interaction and oceanic circulation among other things. GCMs have a spatial resolution comparable to the global synoptic network. The most important use of GCMs lately has been to predict temperature variations resulting from variations in emissions of greenhouse gases [11]. Global circulation models are complicated and complex.

It is important to note that the number of model parameters increases along with the increasing complexity of models. However, many of these parameters are not known with sufficient accuracy. This concerns both radiative forcing and climatic sensitivity.

6.2.1. Solar Irradiance and Global Warming

The most obvious direct effect of solar activity on climate is its effect on the Earth's radiative balance *via* variations in TSI. The 11-year periodicity of TSI has a small amplitude. Moreover, short-term variations of energy input are considerably suppressed by the thermal inertia of oceans. That is why the Schwabe cycle in solar luminosity cannot effectively influence climatic processes, particularly if one takes into account that short-term oscillations of energy input are substantially attenuated by the thermal inertia of oceans, which has a large heat capacity and integrates variations in heat input (see Section 5.2). The attenuation of low-frequency (multidecadal and longer) variations is weaker - the damping factor is 0.75 for a 100-year variation of ΔF [416]. However, it is not known whether TSI actually has such long-term variability. Direct measurements performed using space-born instruments during two or three cycles are not sufficiently long for any decisive conclusion. Nevertheless, many prolonged TSI reconstructions have been compiled using both direct data (sunspot numbers) and proxies (^{14}C, ^{10}Be) (Fig. **6.3**).

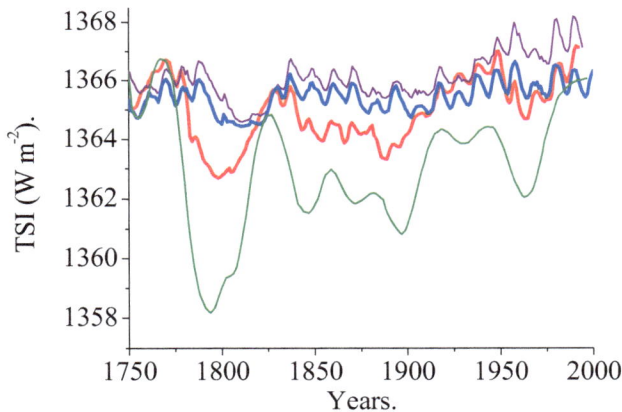

Fig. (6.3). Reconstructions of past variations in the total solar irradiance obtained by: Hoyt and Schatten [564] - red line; Lean *et al.* [565] - purple line; Beer *et al.* [566] - green line, and Mordvinov *et al.* [567] - blue line.

During the 20th century the increase in solar irradiance could have been 1-3 W m^{-2} or more (Fig. **6.3**). According to [568] the mean TSI raised by ca. 4 W m^{-2} during the time interval from the end of the 19th century to the end of the 20th century. That is enough to produce the radiative forcing of $\Delta F_{TSI} = 0.18\text{-}0.70$ W m^{-2}. The IPCC [401] reported a smaller value -an increase in solar irradiance since 1750 is expected to cause a radiative forcing of 0.00-0.10 W m^{-2}. It is easy to estimate the probable temperature response to the solar forcing ΔF_{TSI} over the 20th century. A rough quantitative estimation can be obtained by simple multiplication of radiative forcing by climatic sensitivity λ_c. The values of climatic sensitivity, obtained using different methods including analysis of instrumental temperature data, paleodata, the data of the satellite Earth Radiation Budget Experiment (ERBE) *etc.*, are shown in Table **6.1**.

Table 6.1. Estimations of climate sensitivity.

Author	Method of Estimation	λ_c (°C W^{-1} m^2)
Lindzen and Giannitsis [569]	Climatic response to volcanic explosions (Krakatau, Katmai, Pinatubo)	0.07
Lindzen and Choi [570]	1985 to 1996 ERBE data, sea-surface temperature	0.14
Lindzen *et al.* [571]	Upper-level cloud cover and sea-surface temperature data from the tropical Pacific	0.17-0.43
Douglass *et al.* [572]	Climatic response to Pinatubo explosion	0.22
Schwartz [573]	Global surface temperature	0.25-0.79
Chylek and Lohmann [574]	Global surface temperature, CO_2, aerosol optical depth after 1985	0.29-0.48
Chylek *et al.* [575]	Antarctic paleodata (temperature, CO_2, CH_4, dust) of Late Glacial Maximum to Holocene transition	0.36-0.68
IPCC [401]	Various methods of estimation	0.40-1.22
Forest *et al.* [576]	Global-mean surface (1906-1995) and deep-ocean temperatures (1948-1995)	0.37-2.08
Andronova and Schlesinger [577]	Global mean and hemispheric difference in surface air temperature 1856-1997	0.27-2.54

Available assessments of λ_c - the key parameter of climatic modelling - vary by more than an order of magnitude (Table **6.1**). This difference is a result of the complexity of the climatic system, which has numerous feedbacks. A number of these feedbacks are not well known at present. Considering the assessments of the IPCC [401] we discover that the increase in TSI during the last 100 years could cause the corresponding global warming by 0.07-0.85°C. The difference between the lower and upper limits of the increase in the global temperature induced by the solar luminosity change reaches a factor of twelve. If we take into account the sensitivity estimations not only of the IPCC consortium but also of other authors the difference will be greater. Such uncertainty prevents drawing any decisive conclusions about the solar contribution to global warming. It may be significant as well as negligible. This conclusion agrees well with the conclusion of the IPCC [339] that the modern level of scientific understanding of the TSI-climate relationship is low. It also confirms the fact that our knowledge about the past of solar activity has only a qualitative character.

6.2.2. Greenhouse Gases and Global Warming

According to the IPCC [401] the current level of understanding of greenhouse gas (CO_2, CH_4, N_2O) forcing is high. Anthropogenic radiative forcing from emissions of CO_2 can be assessed by means of a well-known formula (*e.g.* [150]) that approximates fairly accurately the detailed radiative transfer calculation:

$$\Delta F_{CO_2}(x, y) = 5.35 \cdot \ln\left(\frac{[CO_2]_x}{[CO_2]_y}\right) \tag{6.1}$$

where ΔF_{CO2} is in W m^{-2}. Formula (6.1) means that a doubling of $[CO_2]$ can provide a forcing of 3.7 W m^{-2}. An increase in carbon dioxide concentration from 297 ppm in 1900 to 391 ppm in 2011 has caused a forcing of $\Delta F_{CO2}(1900\text{-}2011)$ = 1.6 W m^{-2}. That means that the greenhouse contribution to GW is obviously larger than the direct solar contribution. However, it is not easy to assess the global temperature response to greenhouse forcing due to uncertainty in climatic sensitivity.

Knowledge about other radiative forcing that likely has influenced the terrestrial climate over the last 150 years - (a) human-induced aerosol emissions; (b) human-induced changes in surface albedo (land use, black soot on snow); (c) volcanic-induced change in concentration of aerosol in stratosphere; and (d) aerosol influence on clouds - is still unsatisfactory. As a result, the evaluation of the total net anthropogenic forcing relative AD 1750 ranges from 1.13 W m^{-2} to 3.33 W m^{-2} [401].

Thus, available information about the character and structure of feedbacks is not sufficient [578].

Correspondingly the parameters of many state-of-the-art climate models have large uncertainties. Without entering this discussion, we can note that broad differences in the main model parameters and input data really allows to fit the model-derived temperature to the measured one by a variety of ways. The numerical experiment of Ogurtsov [579] demonstrates that an arbitrary selection of radiative forcing without justification of the choice criteria allows largely explain the global temperature rise till 1970 even without the hypothesis about the greenhouse effect.

6.2.3. Cosmic Rays and Global Warming

Ogurtsov [579] evaluated the climatic consequences of the positive correlation between flux of galactic cosmic rays and low cloudiness claimed by Marsh and Svensmark [459] and Palle and Butler [460]. He considered cloud cover to be affected by GCR intensity and not entirely determined by temperature. Thus cloudiness was regarded as a forcing factor rather than a feedback as assumed traditionally in climatology. Furthermore, this study estimated the possible cloud radiative forcing since the end of the 19th century using the following assumptions: (a) reconstruction of GCR intensity by Mursula *et al.* [580], (b) a linear relationship between low cloud cover and GCR during the period 1983-1994, suggested by Marsh and Svensmark [459], and (c) assessments of cloud radiative forcing using the data of the Earth Radiation Budget Experiment [459]. Using this hypothetical forcing, solar irradiance reconstruction after Hoyt and Schatten [564] (Fig. **6.4B**), and a simple one-dimensional (four latitudinal belts) energy-balance model, Ogurtsov [579] estimated the mean hemispheric temperature during the period 1886-1999. Fig. (**6.4C**) shows that the combined effect of the variations in (a) the TSI and (b) low cloud cover may result in an increase in the mean temperature in the Northern Hemisphere during 20th century by about 0.35°C. Therefore, in the framework of the approach of [579], the warming of the Northern Hemisphere prior to the 1980s could be only due only to variations in solar-cosmic factors, even if the greenhouse forcing is completely neglected.

However, at the end of the 1990s the correlation between GCR and clouds decayed and then changed its sign, which made any conclusions about a cosmic ray-cloud relationship uncertain and questionable (see Section 5.3.2). The estimation of [579] is therefore only a computing exercise, which clearly demonstrates that a subjective handling of the model parameters, neither of which is known exactly enough, can bring us quite questionable results. Taking into account this result as well as several problems remaining in climate modelling [581], it is reasonable to conclude that simulations performed by means of modern climate models still seem to be only fitting of the computation results to the actually observed temperatures. A rather good fit can be ensured using a variety of combinations of input data. Such studies are necessary for analyzing potential

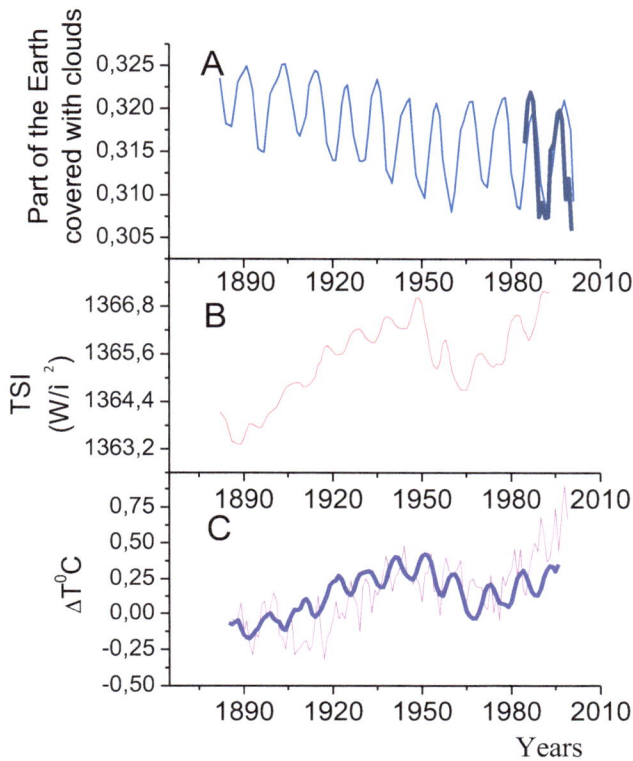

Fig. (6.4). (A) Low cloud cover over the middle latitudes (25-60° N) of the Northern Hemisphere. Thick blue line - experimental data (yearly averages), thin dark blue line - reconstruction from the data on GCR intensity obtained by Mursula *et al.* [580]. **(B)** Reconstruction of the TSI after Hoyt and Schatten [564]. **(C)** Mean annual temperature of the Northern Hemisphere. Thin magenta line -data of weather stations, thick violet line - model calculation [579].

scenarios of climatic variations in the past as well as in the future. Nevertheless it is currently difficult to evaluate (even roughly) the probability of each specific scenario. We can only note that the Sun's contribution to the rapid global warming through the last three to four decades is either minor [582] or negligible [583]. The warming after 1980 was not simulated by the model [579] (see Fig. **6.4C**). This is taken as evidence that the observed fast increase in global mean temperatures since the early 70s is challenging to reproduce in any model without considering the greenhouse forcing. On the other hand, the evaluation of Scafetta [584, 585], who considered the global climate to be appreciably influenced by planetary cycles (particularly the cycles connected with the orbital periods of Jupiter and Saturn), showed that 60% of the temperature increase observed since

1970 could be produced by the joint effect of natural climate oscillations of a celestial origin.

6.2.4. Changes in Background Irradiance and Global Warming

Changes in the global climate of the Earth depend evidently on the background solar irradiance, *i.e.* on the amount of shortwave solar radiation incoming into the atmosphere and the fraction of this radiation that is reflected back to space. It was shown recently that solar radiation incidents at the terrestrial surface increased appreciably during the last decades of the 20th century [586]. This observable fact is often called a global brightening. Estimations of the radiative forcing (Fig. **6.5**) - global perturbation of the radiation balance of the terrestrial atmosphere - are based on the following data:

(a)   Solar irradiance at the Earth's surface, averaged over the globe, which [587] derived from the data on cloudiness in the framework of the International Satellite Cloud Climatology Project (ISCCP) during the period 1983-2001.

(b)   Globally averaged changes in solar irradiance at the Earth's surface during the period 1984-2003, derived from the data on the Earth's reflectance by Palle *et al.* [588] using ISCCP data and the data on dayside earthlight reflected from the Moon.

(c)   Average of the eight records of anomalies in ground-based measurements of solar radiation incident on the Earth's surface during the period 1993-2001 [589].

Changes in surface solar radiation in 1983-2001 caused a positive radiative forcing ranging from 3 W m^{-2} to 6-7 W m^{-2} [586] (Fig. **6.5**). If we take a value of climatic sensitivity adopted by the IPCC ($\lambda_c = 0.40$-1.22 °C W^{-1} m^2), we obtain an increase in global temperature by 1.5-3.6 °C as a result of the radiative forcing of 3 W m^{-2}. Thus, an increase in the amount of solar radiation that reached the terrestrial surface at the end of the 20th century, determined in different ways, should cause a jump in the global temperature, which, however, has not been observed.

Estimation performed by means of the one-dimensional energy-balance climatic model confirmed that the increase of global surface radiation by 3 W× m^{-2} through 1983-2001 should result in a corresponding rise in temperature, which

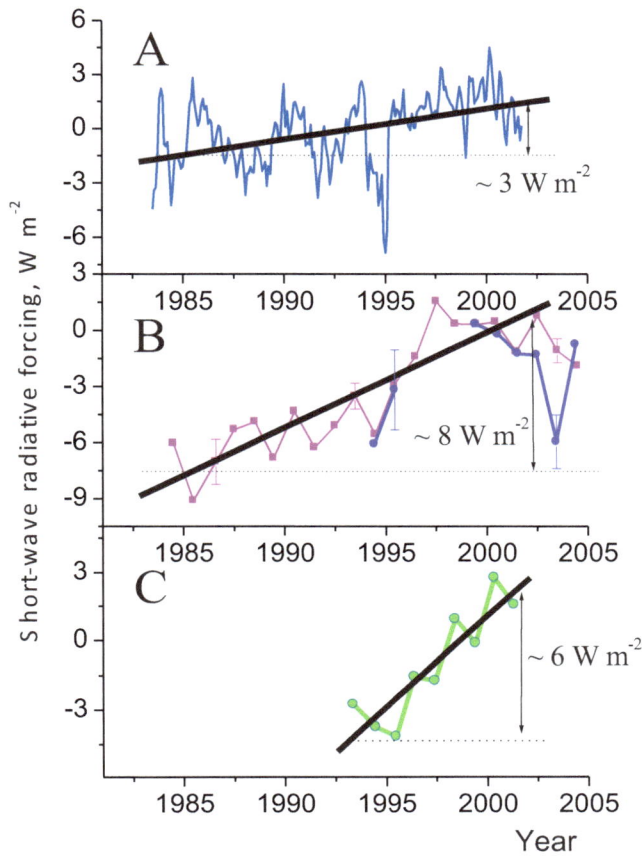

Fig. (6.5). Shortwave radiative forcing assessed by means of: (**A**) satellite-derived surface radiation (the data were electronically scanned and digitized from [587]), (**B**) estimation of the Earth's reflectance made by means of ISCCP data (magenta line with squares) and the data on the Earth's albedo (violet line with circles) [588], and (**C**) surface radiation estimated using the data on ground-based measurement [589]. All the data were electronically scanned and digitized.

exceeds the actual observed values by 0.6°C-2.0°C. Thus, the experimental estimations of solar radiation that reaches the terrestrial surface apparently contradict the actually measured global temperatures at least in the framework of: (a) simplified linear estimations, and (b) the energy balance approach. Evidently, according to [464], the ISCCP data set is not appropriate for long-term trend estimation. On the other hand, global brightening is corroborated by the ground-based observations of the solar radiations and the bright and dark sides of the Moon (earthshine). The inconsistency between the background solar irradiance and global temperature during the period 1983-2001 might be one more indicator of our insufficient knowledge about the climatic system.

6.3. Summary and Remaining Problems

This study of the both instrumental and proxy data on the global temperature, allows us to conclude that the 20th century was warm, *i.e.* the average temperature of the Earth's surface through AD 1900-2000 was indisputably higher than the average temperature during AD 1000-2000. Based on available paleodata we can conclude that the average Northern Hemisphere temperatures in the last two to three decades was:

(a)   undoubtedly highest in the last 500 years.

(b)   likely highest in the last 1000 years. Nevertheless it is hard to show reliably that the last 2-3 decades were the warmest period of the whole last millennium since the existing paleoreconstructions of temperature represented different patterns. This divergence can unlikely be explained by the intrinsic proxy-record uncertainties. (see Section 5.8). Since individual paleo records have insufficient coverage of the Earth's surface the temperature histories based on the paleodata cannot currently be considered to be adequate in their precision or confidence. The forcing factors presumably causing global warming are: (a) anthropogenic greenhouse gases emission, (b) increase in solar activity, (c) the regional anthropogenic impact, and (d) natural climatic cycles. However, it is difficult to evaluate the relative contribution of each of these factors due to lack of knowledge about the respective radiative forcing, climate sensitivity and many other parameters of models. The assessment of total net forcing due to human activity since 1750, made by the IPCC [401], gives a value of 1.13-3.33 $W \times m^{-2}$. Our evaluation of the direct solar forcing, caused by changes in TSI since the beginning of the 19th century, gives a value of $0.18-0.7°W \times m^{-2}$. Thus anthropogenic contribution to GW almost certainly exceeds the direct impact of TSI change. However the Sun can influence the Earth's climate in indirect ways, *e.g.* (a) *via* a relationship between the cloudiness and GCR intensity [459, 529] and (b) *via* a connection between the GCR and SCR flux and aerosol concentration [520]. However, these hypotheses have not been reliably proved and thus it is impossible to obtain quantitative estimation of the corresponding forcing.

Summarizing all the above, we can conclude that the cause of the increase in global temperature over the last 100 years should be considered as not well established due to a lack of detailed knowledge about numerous contributing factors that may be responsible for this phenomenon. The greenhouse effect seems to be a most important contributor to GW, but it is not easy to assess the amplitude of the corresponding temperature rise.

CHAPTER 7

Possible Scenarios of the Solar Activity and Climate Changes in 21ˢᵗ Century

Abstract: A paleoastrophysical forecast of solar activity throughout the 21ˢᵗ century is presented. It is shown that due to our insufficient knowledge of the origin and character of global warming it is hard to predict accurately the next changes in the global climate for this century. Possible scenarios of the future of climate evolution are considered.

Keywords: Climate, prediction, solar activity.

7.1. Projection of Future Evolution of Solar Activity

Prediction of the activity of the Sun is not only a theoretical problem, but it is also hugely important for the very existence of mankind. As solar activity drives a number of physical processes in the Earth's atmosphere and circumterrestrial space, its variability has an effect on many aspects of human livelihood. However, the only option to perform a long-term prediction (at least few decades ahead) is to use paleoastrophysical data.

Ogurtsov [196] used decadal radiocarbon reconstructions of sunspot number over 8555 BC-AD 1605 and predicted an average level of solar activity for the next few decades using both instrumental (AD 1615-2005) and proxy data. The prediction was made by means of a nonlinear approach of Sugihara and May [590], based on the reconstruction of the trajectory of a dynamical system in pseudo-phase space. It was tested using a number of chaotic and random time series [196]. A prognosis of mean decadal Wolf number was obtained for the forthcoming decades (until 2045, Fig. **7.1**).

The accuracy of forecast at each point is evidently rather low (large uncertainty) (Fig. **7.1**). This is not surprising since solar paleoreconstructions probably extract mainly qualitative information about the behaviour of solar activity in the past (see Section 3.8). Nevertheless, we can quite confidently draw a general conclusion for the entire time interval 2015-2045 - in this period SA will plausibly be lower than in the second half of the 20ᵗʰ century. The statistical experiment performed by Ogurtsov [196] with solar paleoproxies showed that the probability of a sunspot with a mean value exceeding 70 (mean value of R_G during

Maxim Ogurtsov, Risto Jalkanen, Markus Lindholm and Svetlana Veretenenko

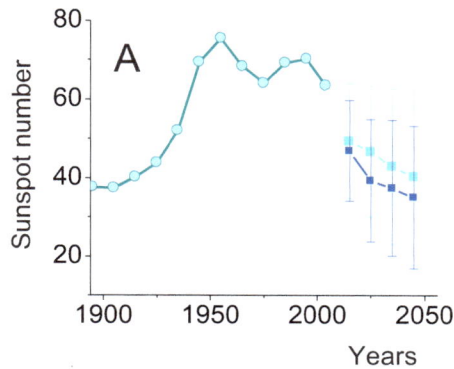

Fig. (7.1). Solar activity prediction. Dark cyan line with circles - real sunspot number averaged over 25 years; cyan line with squares - prediction of sunspot number made using reconstruction after Ogurtsov [196]; blue line with squares - prediction of sunspot number made using reconstruction after Solanki *et al.* [176]. This prediction is based on the dimension of pseudo-phase space d = 3 and maximum distance between neighbours 15.0.

1955-2005) in 2015-2045 is lower than 0.001. This is in line with the prediction made in [176]. At the same time, the probability that the mean value of R_G would be less than 38 was found to be less than 0.1 [196]. These results testify that in spite of the qualitative character of palaeoastrophysical information, long-term (few decades) forecasts of solar activity are possible at least in more general terms (tendency prediction). From the solar paleoastrophysics' point of view the average activity of the Sun would be lower during 2015-2045 than during 1955-2005.

7.2. Projection of Future Evolution of Global Climate

Prediction of future climatic changes is one of the major tasks of climate science. There are two ways to understand the future changes of climate: (a) physical and (b) mathematical. In the former manner of an approach researchers first try to understand the physics of the key climate forcings and describe them with a detailed model. The model is then used to forecast the reaction of future climate to different forcing factors. In the latter manner of an approach researchers take the data of observations and try to extrapolate these appropriately into the future. Attempts of forecasting future global warming started in the 1970s-1980s. These physical predictions were made assuming a mostly greenhouse character of global warming and were based on forecasts of future concentrations of carbon dioxide. The rising concentration of CO_2 has been forecasted rather correctly, *e.g.* [591] forecasted a 375-385 ppm value of CO_2 in the year 2000 while Bolin *et al.* [592] predicted a 375-385 ppm value of the carbon dioxide concentration in the year

2000 while Bolin *et al.* [599] projected a comparable value of 360-380 ppm. Nevertheless, the corresponding prognoses of global surface temperature were not so successful (Table **7.1**).

Table 7.1. Predictions of a mean global temperature in the year 2000 made during 1972-1987. Real temperature in 2000 was 0.39-0.50 T °C.

Source	The Forecast Formula	Predicted Temperature in 2000 (T °C)
Budyko [593]	$T_{2000} = T_{1900} + 1.2$	0.87-1.02
Kellogg [594]	$T_{2000} = T_{1900} + 1.2$	0.87-1.02
Budyko [595]	$T_{2000} = \bar{T}\,(1880\text{-}1975) + 0.6$	0.39-0.48
The impact of atmospheric carbon dioxide increasing on climate [596]	$T_{2000} = T_{1900} + (1.0\text{-}2.0)$	0.67-1.82
Budyko and Izrael [183]	$T_{2000} = T_{1970} + 0.9$	0.82-0.94

Actual global temperatures in the years 1900, 1970 and 2000 were estimated using the data of CRU (http://www.cru.uea.ac.uk/cru/info/warming/) and NCDC (ftp://ftp.ncdc.noaa.gov/pub/data/anomalies/monthly.land_ocean.90S.90N.df_190 1-2000mean.dat). The majority of the forecasts (Table **8.1**) evidently overestimated the temperature rise in the 20th century, because the actual instrumental temperature in 2000 was 0. 39°C (CRU) and 0. 50°C (NCDC). Similarly the detailed prediction made by Hansen *et al.* [597] also failed. These authors carried out three scenarios of potential climate changes based on increasing concentrations of greenhouse gases (CO_2, CH_4, N_2O, *etc.*) till 2020. Scenario A assumed continued exponential rise in greenhouse gases concentration, scenario B assumed a reduced linear growth and scenario C assumed a fast reduction of emissions and thus the total climate forcing stops to increase after the year 2000 (see Fig. **7.2A**). Hansen *et al.* [597] have computed corresponding changes in global temperature for these three scenarios (Fig. **7.2B**). Even the minimum scenario of [597] substantially overestimates (up to 0.1°C) the value of real temperature in 2005-2010. Naturally our knowledge on the climate system has appreciably increased since the end of the 1980s. Nevertheless, the evident failure of the early GW predictions is not encouraging; indicating that more or less precise prognosis of future climate evolution is a very difficult task. Largely inadequate information about the potential forcing factors increases the uncertainty of climatic modelling Current understanding of the past of climate is discrepant, which hinders researchers from reliable and precise extrapolation of temperature into the future. The described shortcomings limit the potential of the both physical and mathematical methods. We can only considered two possible scenarios of global temperature variation in the 21st century. They are based on extreme assumptions about the climate history in the last 100 years.

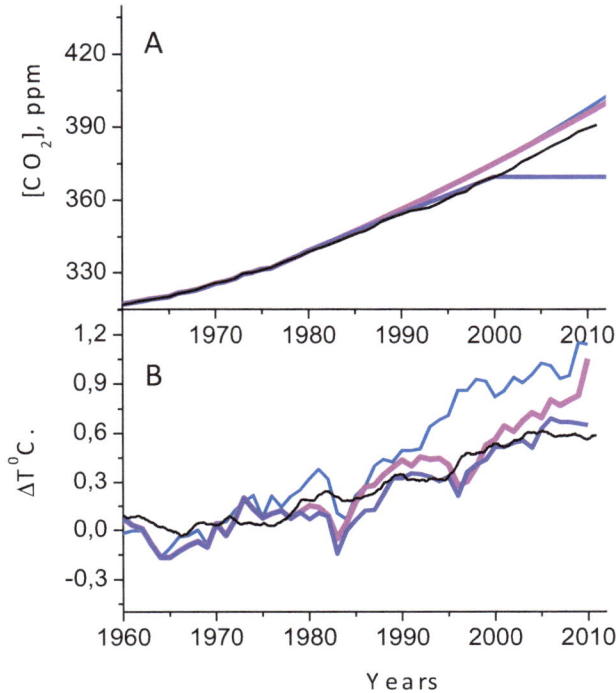

Fig. (7.2). (**A**) Concentration of carbon dioxide in the Earth's atmosphere: black thick line - experimental measurement at Mauna-Loa (www.esrl.noaa.gov/gmd/ccgg/trends/), violet line - scenario C by Hansen *et al.* [597], magenta line - scenario B, blue line - scenario A. (**B**) Global surface temperature calculated for scenarios A, B, and C, compared with instrumental data: black thick line - thermometric measurement (ftp://ftp.ncdc.noaa.gov/pub/data/anomalies/monthly. land_ocean.90S.90N.df_1901-2000mean.dat) smoothed by 55 months, other lines - corresponding scenarios of [597].

(a)   Global warming is extraordinary and unique. It is caused mainly by an additional global-scale forcing factor, which did not work in the past. GW is considered almost completely as a result of the activities of humankind, and the past of the Earth's climatic system does not play significant role (neglect the natural factors). This scenario is in the framework of IPCC paradigm. In that case climatic projections are based on a variety of presumptions on the future development of mankind's emissions of greenhouse gases. They result in a temperature increase of the 21st century ranging from 1.0 to 3.7°C [401].

(b)   Global warming is not unprecedented in a historical context and it is mostly the result of natural (terrestrial, solar-cosmic, astronomic) climatic variations while the influence of greenhouse gases is a minor factor. This scenario of the temperature history of the last millennium agree with MCV and MV paleoreconstructions. Climate of the 20th century was governed by the same dynamic system as in the past (neglect greenhouse effect). In that case the current state of global climate is an expected result of its past, and the paleoclimatic reconstructions (MV and MCV proxy records) could be used as a source of information for predictions. In order to perform a prognosis, we interpolated the paleoreconstructions NHL and NHMb by decades, and made a forecast of mean decadal temperature over the first few decades of the 21st century using a nonlinear prediction approach. The nonlinear forecast was made using the same method as in the solar activity prognosis.

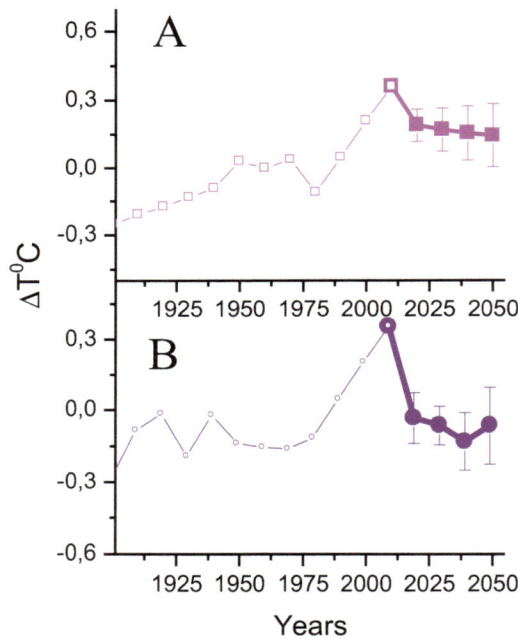

Fig. (7.3). Predictions of future Northern Hemisphere temperatures. (**A**) Forecast based on reconstruction by Loehle [364]. (**B**) Forecast based on the reconstruction by Moberg *et al.* [363]. Both reconstructions were extrapolated till 2010 by means of the instrumental temperature data. The original data, averaged by 13 years and interpolated by decades, are shown with thin lines. Forecasted values are shown with thick lines. This prognosis was made using the embedding dimension d = 3 and ten nearest neighbours.

One can see from Fig. (**7.3**) that the uncertainty of the prediction is large and therefore the reliability of the forecast is low. This is a result of the non-linear and maybe chaotic character of the Earth's climatic system. However, even in that case we can assess the probability of higher temperatures in the first part of the current century. A statistical experiment, performed using the predicted temperature values and the forecast error, showed that this probability is less than 0.25 This means that if the global warming of the last century is a result of natural climatic variability, *i.e.* the modern climate is driven by the same dynamic system that operated in the, the mean temperature of the Northern Hemisphere in the first part of the 21st century will likely be lower than the current value (Fig. **7.3**) [386]. This is in line with the prediction made by Scafetta [585].

If the global warming is a result of a number of climatic processes of both human-induced and natural origin, neither of which could be neglected (industrial greenhouse gas emissions, changes in solar activity, inherent variability of the climate system, regional and local anthropogenic effects), the situation is most intricate. In that case it is not easy to make even a rough evaluation of variations of global climate during current century due to the considerable uncertainty in our information on the relative contributions of the specified forcings (Fig. **7.4**).

7.3. Summary and Prospects for Further Research

The huge progress has been achieved in helioclimatology over the past few decades. Both space-born and ground-based observations have brought a lot of new evidence for a connection between the Sun's activity and the phenomena of the lower atmosphere. The progress in theory, experiment and modelling has also considerably increased our knowledge of the Sun and solar-terrestrial connections. The field, which started as an examination of statistical relationships has developed to a significant branch of science that combines studies of solar and atmospheric physics, climatology, paleoclimatology and paleoastrophysics. Nevertheless, the absolutely conclusive and unequivocal proof of the reality of a solar-climate link is still missing. The lack of scientific facts and understanding about the connecting processes at work is the main cause of this shortcoming. Data obtained by experimental observation are quite precise but rather short in time scale as well as scarce in distribution. Paleoproxies are much longer but their uncertainty generally increases as a function of time from the present. Thus, in solar-climatic research we face a kind of 'uncertainty principle':

$$\frac{\Delta X}{\Delta T} \approx \text{const,} \qquad\qquad (7.1)$$

where ΔX is the uncertainty of the data, and ΔT is the length of the X time series.

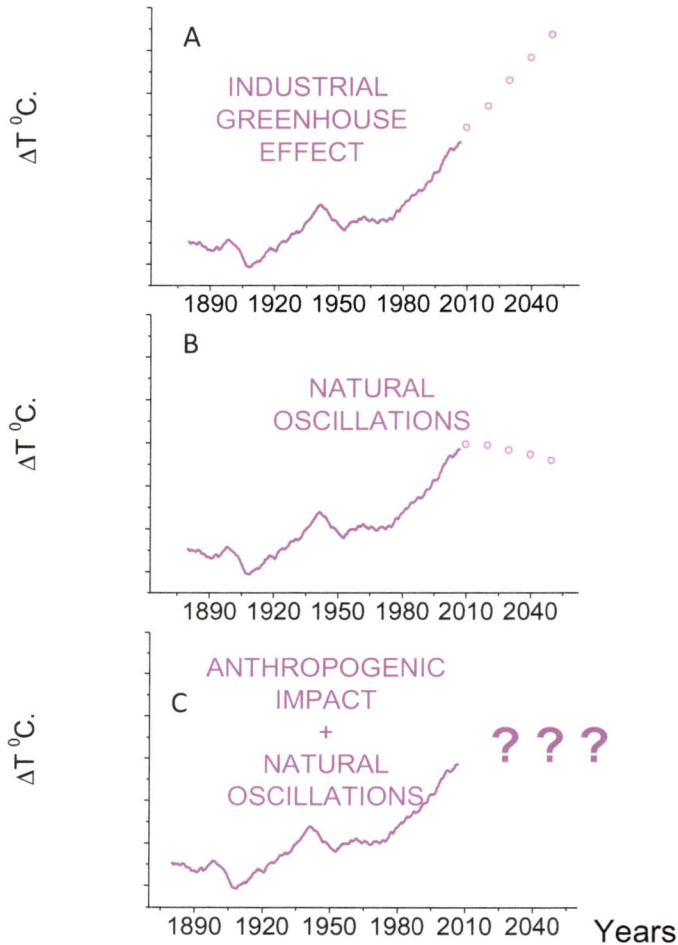

Fig. (7.4). Possible scenarios of the evolution of climate in the 21st century. (**A**) global warming is the result of man-made emission of greenhouse gases; (**B**) global warming is a result of natural climatic oscillations; (**C**) global warming is a result of a complex of anthropogenic and natural forcings.

Solar paleoreconstructions contain only qualitative (not quantitative) information of the past of the Sun's activity and climatic paleoproxies often are even more uncertain. Despite the considerable successes achieved by paleoclimatology, its methods and approaches still leave appreciable space to subjectivity. Moreover, it has been shown that the Sun-climate connection, even if it actually exists, may be realized by an indirect and nonlinear way. In addition, the solar- driven variations of climate can be distorted by natural fluctuations not linked to the activity of the Sun. As a result the search for a connection between the Sun's activity and weather and climate has turned out to be a more demanding task than was previously anticipated. Analyses of all available climatic records (both direct data and proxies) show that global temperature of the last 20-30 years was warmer than at any corresponding time interval over the last 500 years. The forcing factors, which most likely caused the global warming are: (a) human-caused variations in the atmospheric concentration of greenhouse gases and aerosols, (b) variations of activity of the Sun, (c) changes in activity of volcanoes, (d) internal oscillations in the climatic system and (e) human-induced variations in land surface properties. Greenhouse forcing is often considered as the key contributor to global warming. Climatic modeling gives evidence in favor of this hypothesis, because it is difficult to simulate a sharp increase in temperature starting in the 1970s, if greenhouse forcing is neglected. However, it should be noted that a lack in adequate understanding about the parameters and particularly feedbacks of climatic models makes the results of modelling rather ambiguous. Moreover, paleoclimatic data do not strongly support the greenhouse hypothesis. Indeed, studies of ice-core records spanning the last 9-420 kA disclose a pattern of strong increases in temperature and carbon dioxide concentration at ca 100,000-year time intervals. But the CO_2 increase usually has come 400-1000 years after (not before) the temperature rise [598, 600]. This relationship probably appears as higher temperatures have facilitated a release of the carbon dioxide from oceans.

Available information shows that the assumption of greenhouse character of the global warming (often considered the prevailing paradigm in climatology) in fact is merely a scientific hypothesis. This hypothesis looks plausible and actually helps in explaining the considerable increase in global surface temperature in the last 100 years, but it still needs serious additional substantiation. Future work is here envisaged to require improvement in our understanding of both solar-climate relationship as well as the principles governing long-term climatic evolution:

1) Further gathering and accumulation of data and information together with its subsequent systematization. This concerns both instrumental monitoring of solar-cosmic and geophysical parameters and work on constructing new proxies (paleoreconstructions) of solar activity and climate. The development of methods of paleoastrophysics and paleoclimatology is an integral part of this work.

2)   Improvement of the methods of statistical analysis, particularly methods aimed at the search and detection of nonlinear interrelations between the different time series.

3)   Further laboratory research of the physical processes which possibly provide a relationship between activity of the Sun and the low atmosphere.

4)   Increasing our understanding about the climatic system and improvements of methods of climatic modelling. Currently it seems dubious that the problem of the origin and sources of contemporary rise in global temperature will be solved before we reach the mid-21st century.

List of Abbreviations

ARS	=	Anomalous reduction in the sensitivity
CCN	=	Cloud condensation nuclei
CME	=	Coronal mass ejections
CRIF	=	Cosmic ray induced freezing
D-O	=	Dansgaard-Oeschger events
FD	=	Forbush decrease
GCR	=	Galactic cosmic ray
IHS	=	Ice-hockey stick
IMF	=	Interplanetary magnetic field
IPCC	=	Intergovernmental panel on climate change
LIA	=	Little ice age
MCV	=	Multi-centennial variability.
MV	=	Millennial variability
MWP	=	Medieval warm period
PCA	=	Polar cap absorptions
PSCs	=	Polar stratospheric clouds
SCR	=	Solar cosmic rays
RCS	=	Regional curve standardization
SCL	=	Solar cycle length
SONE	=	Sunspots observed by naked eye
SPE	=	Solar proton event
VEI	=	Volcanic explosivity index

Appendices

APPENDIX 1

Calculation of the Geomagnetic Cutoff

In order to estimate the geomagnetic cut-off and to determine the energy of particles which can penetrate into the Earth's atmosphere at a given geographic point with coordinates φ and λ, it is reasonable first to calculate the geomagnetic coordinates of the point. Geomagnetic latitude Φ and longitude Λ can be calculated using transformations [2]:

$$\sin \Phi = \sin \varphi \cdot \sin \varphi_0 + \cos \varphi \cdot \cos \varphi_0 \cdot \cos(\lambda - \lambda_0),$$
$$\sin \Lambda = \frac{\cos \varphi \cdot \sin(\lambda - \lambda_0)}{\cos \Phi}, \tag{A1.1}$$

where $\phi_0 = 78.3°$ N, $\lambda_0 = 291°$ E ($\phi_0 = 78.3°$ S, $\lambda_0 = 111°$ E) - modern geographic coordinates of the northern (southern) magnetic pole. In order to hit the Earth at geomagnetic latitude Φ the particle must have a rigidity of more than R_c:

$$R_c = 14.9 \cdot \cos^4 \Phi, \tag{A1.2}$$

where R_c is in GV. Calculation results in the following values: $R_c = 0.53$ GV at Murmansk ($\phi=69°$, $\lambda=33°$), $R_c = 0.61$ GV at northern Fennoscandia ($\phi=67°$, $\lambda=27°$), $R_c = 1.43$ GV at Saint-Petersburg ($\phi=60°$, $\lambda=27°$), and $R_c = 2.28$ GV at Moscow ($\phi=56°$, $\lambda=37°$). The precise values of R_c are 0.6 GV (Murmansk), 1.7 GV (St. Petersburg), and 2.4 GV (Moscow) [527]. Therefore formula (A1.2) is rather suitable for rough estimations. The energy of a proton corresponding to the calculated rigidity can be determined using a diagram in Fig. (**A1.1**).

Fig. (**A1.1**) shows that at the Kola Peninsula and in northern Fennoscandia protons with energies 130-170 MeV can penetrate the atmosphere. It should be taken into account that during strong geomagnetic storms geomagnetic cut-off could decrease by 50-80% or more [74]. For example, at the Dst minimum on 20 November, 2003 geomagnetic cut-off at Moscow dropped to 0.3 GV and thus was lower than in the Apatity station, which is situated close to the auroral oval and Murmansk.

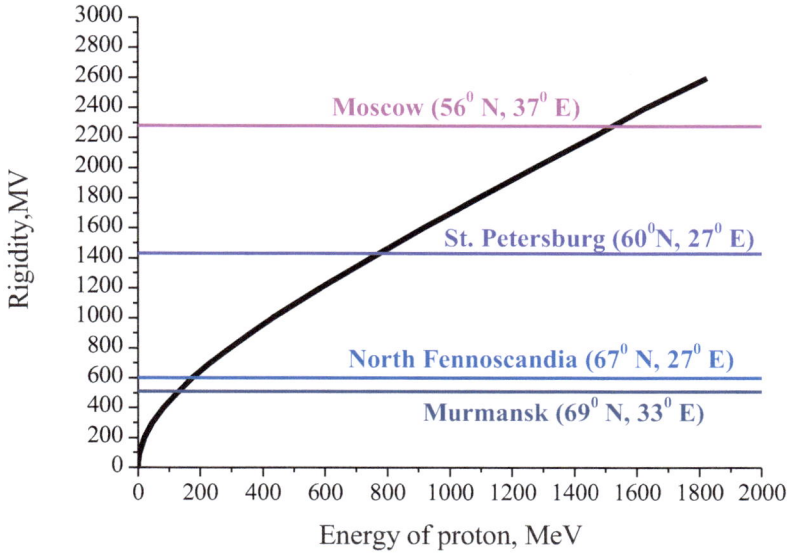

Fig. (A1.1). Dependence of proton rigidity on its energy.

APPENDIX 2

Some Important Characteristics of the Atmosphere

Altitude, m	Temperature, K	Pressure, Pa	Density kg×m^{-3}	Dynamic Viscosity, 10^{-4} g×s^{-1}×cm^{-2}	Number Density of Air Molecules, cm^{-3}	Aerosol Optical Depth
0	288.2	1013.3	1.225	1.79	2.55(19)	0.125
1000	281.7	898.8	1.112	1.76	2.35(19)	0.083
2000	275.2	795.0	1.006	1.72	2.14(19)	0.058
3000	268.7	701.2	0.909	1.69	1.94(19)	0.042
4000	262.2	616.6	0.819	1.66	1.73(19)	0.034
5000	255.7	540.5	0.736	1.63	1.53(19)	0.028
6000	249.2	472.2	0.660	1.60	1.40(19)	0.024
7000	242.7	411.1	0.590	1.56	1.26(19)	0.021
8000	236.2	356.5	0.526	1.52	1.13(19)	0.017
9000	229.7	308.0	0.467	1.49	9.94(18)	0.014
10000	223.3	265.0	0.413	1.46	8.60(18)	0.011
11000	216.8	227.0	0.365	1.42	7.69(18)	0.008
12000	216.7	194.0	0.312	1.42	6.78(18)	5.0(-3)

(Appendix 2) contd…..

Altitude, m	Temperature, K	Pressure, Pa	Density kg×m^{-3}	Dynamic Viscosity, 10^{-4} g×s^{-1}×cm^{-2}	Number Density of Air Molecules, cm^{-3}	Aerosol Optical Depth
14000	216.7	141.7	0.228	1.42	4.96(18)	3.9(-3)
16000	216.7	103.5	0.166	1.42	3.61(18)	2.8(-3)
18000	216.7	75.6	0.122	1.42	2.73(18)	1.8(-3)
20000	216.7	55.3	0.089	1.42	1.85(18)	1.15(-3)
24000	220.6	29.7	0.047	1.44	1.03(18)	4.2(-4)
28000	224.5	16.2	0.025	1.47	5.63(17)	1.6(-4)
32000	228.5	8.9	0.014	1.50	3.00(17)	5.4(-5)
36000	239.3	5.0	0.007	1.55	1.57(17)	1.6(-5)
40000	250.4	2.9	0.004	1.60	8.31(16)	5.3(-6)
50000	270.7	0.8	0.001	1.71	2.13(16)	0.0
60000	247.0	0.22	0.0003	1.58	4.70(15)	0.0

References

[1]   Khrgian AKh. Physics of the atmosphere. Leningrad 1969; 647 p (in Russian).

[2]   Kallenrode M-B. Space Physics: an introduction to plasmas and particles in the heliosphere and magnetospheres (third edition). Springer-Verlag 2004; p. 482.

[3]   Ogurtsov M, Lindholm M, Jalkanen R. Solar activity, space weather and the Earth's climate. In: Hannachi A, Ed. Climate variability - some aspects, challenges and prospects. Rieka: InTech 2012; pp. 39-72.

[4]   Frankighoul C, Hasselmann K. Stochastic climate models. Part 2, Application to sea-surface temperature anomalies and thermocline variability. Tellus 1977; 29: 289-305.

[5]   Gold T. Plasma and magnetic fields in the solar system. J Geophys Res 1959; 64: 1665-1674.

[6]   Cannon PS. Extreme space weather - a report published by the UK Royal Academy of Engineering. Space Weather 2013; 11: 138-139.

[7]   Daglis I, Baker D, Kappenman J, *et al*. Effects of space weather on technology infrastructure. Space Weather 2004; 2: S02 004.

[8]   Schrijver C, Mitchell S. Disturbances in the US electric grid associated with geomagnetic activity. J Space Weather Space Clim. 2013; 3: A19.

[9]   Siscoe G. The space weather enterprise: Past, present, and future. J Atmos Solar-Terr Phys 2000; 62: 1223-1232

[10]   van der Plicht J, Jull AJT. Mammoth extinction and radiation dose: a comment. Radiocarbon 2011; 53: 713-715.

[11]   IPCC, WGI Fourth Assessment Report: Climate Change 2007: The Physical Science Basis: Summary for Policymakers. Paris 2007.

[12]   Khromov SP 1973. On some doubtful questions related to the solar activity cyclicity and its possible connections with climate. Meteorology and Hydrology 1973; 9: 93-99 (in Russian).

[13]   Pittock AB. A critical look at long-term sun-weather relationships. Rev Geophys Space Phys 1978; 16: 400-420.

[14]   Pittock AB. Solar variability, weather and climate: an update. Quart J Roy Met Soc 1983; 109: 23-57.

[15]   Pittock B. Can solar variations explain variations in the Earth's climate? Clim Change 2009; 96: 483-487.

[16]   Obridko VN. Sunspots and active complexes Moscow: Nauka 1985; 256 p. (in Russian).

[17]   Obridko VN. Magnetic fields and active complexes. In: Zelenyi LM, Veselovsky IS, Eds. Plasma HelioGeoPhysics monograph. Moscow: Fizmatlit 2008; pp. 41-60 (in Russian).

[18]   Solanki S. Sunspots: An overview. Astron Astrophys Rev 2003; 11: 153-286.

[19]   Ringnes TS, Jansen E. On the relation between magnetic fields and areas of sunspots in the interval 1917-56. Astrophys Norvegica 1960; 7: 99-121.

[20]   Solov'ev AA, Kirichek EA. Solar activity and its magnetic origin. In: Bothmer V, Hady AA, Eds. Proc IAU Symposium no. 233. Cambridge Univ Press 2006; pp. 523-524.

[21]   Priest ER. Solar flare magnetohydrodynamics. New York: Gordon and Breach Science Publishers 1981.

[22]   Shibata K, Isobe H, Hillier A, *et al*. Can superflares occur on our Sun ? Proc Astr Soc Japan 2013; 65: 49-1 - 49-8.

[23]   Stix M. The Sun. An introduction. Springer 2002; p. 92.

[24]   Stephenson FR. Historical Evidence concerning the Sun: Interpretation of sunspot records during the telescopic and pretelescopic eras. Phil Trans Roy Soc Lond 1990; A330: 499-503.

[25]   Wittmann AD, Xu Z. A catalogue of sunspot observations from 165 BC to AD 1684. Astron Astroph Suppl Ser 1987; 70: 83-94.

[26]   Vaquero JM. Historical sunspot observations: A review. Adv Space Res 2007; 40: 929-941.

[27]   O'Dell CR, Van Helden A. How accurate were the seventeenth-century measurements of solar diameter. Nature 1987; 330: 629-631.

[28]   Legrand JP, Le Goff M, Mazaudier C, Schroder W. Solar and auroral activities during the seventeenth century. Acta Geod Geoph Mont Hung 1992; 27: 251-282.

[29]   Eddy A. The Maunder minimum. Science 1976; 192: 1189-1202.

[30]   Ribes JC, Nesme-Ribes E. The solar sunspot cycle in the Maunder minimum AD 1645 to AD 1715. Astron Astrophys 1993; 276: 549-563.

[31]   Hoyt D, Schatten KH. How well was the sun observed during the Maunder minimum? Solar Phys 1996; 165: 181-192.

[32]   Legrand MR, Delmas RJ. Relative contributions of tropospheric and stratospheric sources to nitrate in Antarctic snow. Tellus1986; 38B: 236-249

[33]   Vaquero JM, Trigo RM, Gallego MC, Moreno-Corral MA. Two early sunspots observers: Teodoro de Almeida and Jose Antonio Alzate. Solar Phys 2007A; 240: 165-175.

[34]   Usoskin IG. A history of solar activity over millennia. Living Rev Solar Phys 2008; 5: 3-87.

[35]   Vitinsky YuI. The solar cyclicity. Moscow: Nauka 1969; p. 92. (in Russian).

[36]   Usoskin IG, Mursula K, Kovaltsov GA., Lost sunspot cycle in the beginning of Dalton minimum: New evidence and consequences. Geophys Res Lett 2002a; 29: 36-1.

[37]   Hoyt D, Schatten KH. Group sunspot numbers: a new solar activity reconstruction. Solar Phys 1998; 179: 189-219.

[38]   Waldmeier M. The sunspot activity in the years 1610-1960. Zurich: Schulthess 1961.

[39]   Nagovitsyn YuA. To the description of long-term variations in the solar magnetic flux: the sunspot area index. Astron Lett 2005; 31: 557-562.

[40]   Vaquero JM, Gallego MC, Trigo RM. Sunspot numbers during 1736-1739 revised. Adv Space Res 2007B; 40: 1895-1903.

[41]   Ogurtsov MG. Instrumental data on the sunspot formation in the 17th-18th centuries: correct information or approximations. Geomagn Aeron 2013; 53: 663-671.

[42]   Nagovitsyn YuA, Ivanov VG, Miletsky EV, Volobuev DM. ESAI data base and some properties of solar activity in the past. Solar Phys 2004; 224: 103-112.

[43]   Heath DF, Schlesinger BM. The Mg 280-nm doublet as a monitor of changes in solar ultraviolet irradiance. J Geophys Res 1986; 91: 8672-8682

[44]   Dudok de Vit T., Kretzschmar M., Lilensten J., Woods T. Finding the best proxies for the solar UV irradiance. Geophys Res Lett 2009;36L10107,doi:10.1029/2009GL037825.

[45]   Kleczek J. Solar flare index calculations. Publ Centr Astron Inst Czechoslovakia 1952; 22.

[46]   Willson RC, Mordvinov AV. Secular total solar irradiance trend during solar cycles 21-23. Geophys Res Lett 2003; 30: 1199-1202. doi:10.1029/2002GL016038.

[47]   Fröhlich C, Lean J. The Sun's total irradiance: cycles and trends in the past two decades and associated climate change uncertainties. Geophys Res Lett 1998; 25: 4377- 4380.

[48]   Stewart JQ, Panofsky AA. The mathematical characteristics of sunspot numbers. Astrophys J 1932; 88: 385-391.

[49]   Gleissberg W. A long-periodic fluctuation of the sun-spot numbers. Observatory 1939; 62:158-159.

[50]   Gleissberg W. A preliminary forecast of solar activity. Popular Astron 1948; 56: 399-402

[51]   Nagovitsyn YuA. Solar activity during the last two millennia: solar patrol in Ancient and Medieval China. Geomagn Aeron 2001; 41: 680-688.

[52]   Ogurtsov MG, Nagovitsyn YA, Kocharov GE, Jungner H. Long-period cycles of the Sun's activity recorded in direct solar data and proxies. Solar Phys 2002; 211: 371-394.

[53]   Vitinsky YI, Kopecky M, Kuklin GV. Statistics of sunspot activity. Moscow; Nauka1986 (in Russian).

[54]   Hathaway DH, Wilson RM, Reichman ES. Group sunspot numbers: sunspot cycle characteristics. Solar Phys 2002; 211: 357-370.

[55]   Gnevyshev MN, Ohl AI. On the 22-year cycle in solar activity. Astr. Reports 1948; 25: 18-20 (in Russian).

[56]   Waldmeier M. Astron Mitt Zurich 1935; 138: 439.

[57]   Veselovski IS, Tarsina MV. Intrinsic nonlinearity of the solar cycles. Adv Space Res 2002; 29: 417-420.

[58]   Karak BB, Choudhuri AR. The Waldmeier effect and the flux transport solar dynamo // Month Not Roy Astron Soc; 410: 1503-1512.

[59]   Chernosky EJ. A relationship between length and activity of sunspot cycles. Publ Astron Soc Pacific 1954; 66: 241-245.

[60]   Wilson RM, Hathaway DH, Reichmann EJ. An estimate for the size of cycle 23 based on near minimum conditions. J Geophys Res 1998; 103: 6595-6603.

[61]	Dicke RH. Is there a chronometer hidden deep in the Sun? Nature 1978; 276: 676-680.
[62]	Friis-Christensen E, Lassen K. Length of solar cycle: an indicator of solar activity closely associated with climate. Science 1991; 254: 698-700.
[63]	Ogurtsov M, Jungner H. Temporal evolution of statistical features of the sunspot cycles. Adv Space Res 2012; 50: 669-675.
[64]	Maunder E Walter. A prolonged sunspot minimum. Knowledge 1894; 17: 173-176.
[65]	Maunder E Walter. The prolonged sunspot minimum, 1645-1715. J Brit Astronom Soc 1922; 32: 140, 1922
[66]	Eddy JA, Stephenson FR, Yau KKC. On pre-telescopic sunspot records. Royal Astron Soc Quatern J 1989; 30: 65-69.
[67]	Pulkkinen T. Space weather: terrestrial perspective. Living Rev. Solar Phys 2007; 4: 1-60.
[68]	Svalgaard L, Cliver EW. The IDV index: Its derivation and use in inferring long-term variations of the interplanetary magnetic field strength. J Geophys Res 2005; 110: A12103, doi:10.1029/2005JA011203.
[69]	Illing RME, Hundhausen AJ. Disruption of a coronal streamer by an eruptive prominence and coronal mass ejection. J Geophys Res 1986; 91: 10951-10960.
[70]	Bothmer V, Schwenn R. The structure and origin of magnetic clouds in the solar wind. Ann Geophysicae 1998; 16: 1-24.
[71]	Burlaga LF. Magnetic clouds and force-free fields with constant alpha. J Geophys Res 1988; 93: 7217-7224.
[72]	Vandas M, Romashets EP. A force-free field with constant alpha in an oblate cylinder: A generalization of the Lundquist solution. Astron Astrophys 2003; 398: 801-807.
[73]	Burlaga, L. F., R. M. Skoug, C. W. Smith, D. F. Webb, T. H. Zurbuchen, and A. Reinard (2001), Fast ejecta during the ascending phase of solar cycle 23 ACE observations, 1998- 1999, J. Geophys. Res., 106, 20,957-20,977
[74]	Gopalswamy N, Akiyama S, Yashiro S, Makela P. Coronal mass ejections from sunspot and non-sunspot regions. In: Hasan SS, Rutten RJ, Eds. Magnetic coupling between the interior and atmosphere of the Sun. Berlin: Springer-Verlag 2009; pp. 289-308.
[75]	Tyasto M I, Danilova OA, Vernova ES, *et al.* Effect of strongly perturbed magnetosphere on the cosmic-ray cutoff rigidity. Bull RAS: Physics 2007; 71: 1000-1002.
[76]	Goncharova MYu, Maltsev YuP. Correlation of the Kp index with the solar wind parameter. Geomagn Aeron 2001; 41: 317-321.
[77]	Schröder W. Some aspects on the earlier history of solar-terrestrial physics. Wilfried Schröder/Science Edition 2004; 151 p.
[78]	Kurt V, Belov A, Mavromichalaki H, Gerontidou M. Statistical analysis of solar proton events. Ann Geophys 2004; 22: 2255-2271, doi:10.5194/angeo-22-2255-2004.
[79]	Belov AV, Eroshenko EA, Kryakunova ON, *et al.* Ground level enhancements of solar cosmic rays during the last three solar cycles. Geomag. Aeronom 2010; 50: 21-33.
[80]	Belov A, Eroshenko E, Mavromichalaki H, *et al.* A study of the ground level enhancement of 23 February 1956. Advances Space Res 2005; 35: 697-701.
[81]	Bieber JW, Clem J, Evenson P, *et al.* Largest GLE in half a century: neutron monitor observations of the January 20, 2005 event. In: Proceedings of the 29th International Cosmic Ray Conference. Pune, India 1 2005; pp. 237-240.
[82]	Yuqian M. L3 Collaboration: 2003. Search for a muon flux enhancement during the solar flare of 14 July 2000 with the L3+C data. In: Proc. of the 28th International Cosmic Ray Conference. July 31-August 7, 2003. Tsukuba, Japan; p. 3393-3396.
[83]	Miroshnichenko L. Some astrophysical aspects in the studies of solar cosmic rays. Proc. of the 28th International Cosmic Ray Conference. July 31-August 7, 2003. Tsukuba, Japan. 2003; pp. 3355-3361.
[84]	Avakyan SV, Vdovin AI, Pustarnakov VF. Ionizing and Penetrating Radiations in Near-Earth-Space. A Handbook. Petersburg: Hydrometeoizdat 1994; 503 p.
[85]	Ohl AI. Manifestations of solar activity in magnetosphere and ionosphere of the Earth. In: Influence of solar activity on the terrestrial magnetosphere and biosphere. Moscow: Nauka 1971; pp. 104-125 (in Russian).
[86]	Schwenn R. Space Weather: The Solar Perspective. Living Rev Solar Phys 2006; 3: 2-76.

[87]   Galper AM, Gratchev VM, Dmitrenko VV, *et al.* New high energy electron component of Earth radiation belt - high energy electrons. JETP Letters 1983; 38: 409-411 (in Russian).

[88]   Makhmutov VS, Bazilevskaya GA, Stozhkov YI, *et al.* Energetic electron precipitation events recorded in the Earth's polar atmosphere. Proc 32nd International Cosmic Ray Conference. Doi: 10.7529/ICRC2011/V06/0299 2011.

[89]   Stozhkov YuI, Svirzhevsky NS, Bazilevskaya GA, *et al.* Adv Space Res 2009; 44: 1124-1137

[90]   Yermolaev Yu I, Yermolaev MYu. Statistical relationships between solar, interplanetary, and geomagnetospheric disturbances, 1976-2000. 2. Cosmic Research 2003: 41: 115-123.

[91]   Yermolaev YuI, Yermolaev MYu. Statistic study of the geomagnetic storm effectivness of solar and interplanetary events. Adv Space Res 2006; 37: 1175-1181.

[92]   Dorman LI. Cosmic Rays in the Earth's Atmosphere and Underground. Kluwer Ac Publ. Dordrecht/Boston/London, 2004.

[93]   Kudela K, Storini M, Hofer MY, Belov A. Cosmic rays in relation to space weather. Space Science Rev 2000; 93: 153-174.

[94]   Jokipii JR., Variation of the cosmic-ray flux in time. In: ed. by Sonnett CP, Giampapa MS, Matthews MS, Eds. The Sun in time. Tuscon: University of Arizona press 1991, pp. 205-220.

[95]   Stozhkov YuI., Pokrevsky PE, Zullo J Jr, *et al.* Influence of charged particle fluxes on precipitation. Geomagn Aeron 1996; 36: 211-216 (in Russian).

[96]   Stozhkov YI. The role of cosmic rays in the atmospheric processes. J Phys. G: Nucl Part Phys 2003; 29: 913-923.

[97]   Ogurtsov MG, Jungner H, Kocharov GE, *et al.* On the link between northern Fennoscandian climate and length of quasi eleven-year cycle in galactic cosmic-ray flux. Solar Phys 2003; 218: 345-357.

[98]   Kocharov GE, Ostryakov VM, Peristykh AN, Vasil'ev VA. Radiocarbon content variations and Maunder minimum of solar activity. Sol Phys 1995; 159: 381-395.

[99]   Mishra AP, Mishra BN, Gupta M, Mishra VK. 2008. Heliographis distribution of bright solar flares and association of Forbush-decreases with flares and coronal mass ejections. Indian Journal of Radio and Space Physics, V. 37, pp. 237-243

[100]   Jackman HC. Energetic particle influences on NOy and ozone in the middle atmosphere. Geophys Monogr 75 IUGG 1993; 15: 131-139.

[101]   Krivolutsky AA, Repnev AI. Cosmic influences on the ozonosphere of the Earth. Moscow: GEOS 2009; 384 p.

[102]   Reid GC. Associative detachment in the mesosphere and the diurnal variation of polar cap absorption. Plan. Space Science 1969; 17: 731-736.

[103]   Vaquero JM, Vázquez M. The Sun recorded through history: scientific data extracted from historical documents. Astrophys Space Sci Library 2009; 361. Berlin: Springer.

[104]   Schröder W, Shefov NN, Treder H-T. Estimation of past solar and upper atmosphere conditions from historical and modern auroral observations. Annales Geophisicae 2004; 22: 2273-2276.

[105]   Panasyuk MI. Break into the space. Science in Russia 2008; 4: 61-66 (in Russian).

[106]   Gringauz KI, Bezrukhih VV, Ozerov VD, Rybchinsky RE. Research of the interplanetary ionized gas, energetic electrons and corpuscular solar radiation by means of the three-electrode traps of charged particles installed onboard of the second Soviet space rocket. Docl Acad Nauk 1960; 131: 1301-1312 (in Russian).

[107]   Chizhevsky AL. Terrestrial echo of solar storms. Moscow: Mysl'1973, 349 pp. (in Russian).

[108]   Wild G. On the air temperature in Russian empire. St. Petersburg 1882; 2: 765 p. (in Russian).

[109]   Ney EP. Cosmic radiation and weather. Nature 1959; 183: 451-452.

[110]   Dickinson E. Solar variability and the lower atmosphere. Bull Amer Meteorol Soc 1975; 56: 1240-1248.

[111]   Torrence C, Compo GP.: A practical guide to wavelet analyses. Bull American meteorol soc 1998; 79: 61-78.

[112]   Astafieva N M. Wavelet analysis: basic theory and some applications. Phys-Usp 1996; 39: 1085-1108.

[113]   Grossman A, Morlet J.: Decomposition of hardy functions into square integrable wavelets of constant shape. SIAM J Math Anal 1984; 15: 723-726.

[114]   Mallatt SG. Multiresolution approximations and wavelet orthonormal bases of L^2(R). Trans American Math Soc 1989; 315: 69-71.

[115]   Percival DB. On estimation of the wavelet variance. Biometrika 1995; 82: 619-631.
[116]   Beer J, Blinov AV, Bonani G *et al*. Use of Be-10 in polar ice to trace the 11-year cycle of solar activity. Nature 1990; 347: 164-166.
[117]   Berggren A-M, Beer J, Possnert G., *et al*. A 600-year annual ^{10}Be record from the NGRIP ice core, Greenland. Geophys Res Lett 2009; 36: L11801, doi:10.1029/2009GL038004.
[118]   Hasselmann K. Stochastic climate models. Part I. Theory. Tellus 1976; 28: 73-486.
[119]   Wigley TML, Raper SCB. Natural variability of the climate system and detection of the greenhouse effect. Nature1990; 344: 324-327.
[120]   Ogurtsov MG, Kocharov GE, Lindholm M, *et al*. Evidence of solar variation in tree-ring-based climate reconstructions. Solar Phys 2002; 205: 403-417.
[121]   Mudelsee M. TAUEST: a computer program for estimating persistence in unevenly spaced weather/climate time series. Computers Geosciences 2002; 28: 69-72.
[122]   Schulz M, Mudelsee M. REDFIT: estimating red-noise spectra directly from unevenly spaced paleoclimatic time series. Computers Geosciences 2002; 28: 421-426.
[123]   Horne JH, Baliunas SL. A prescription for period analysis of unevenly spaced time series. Astrophys J 1986; 302: 757-763.
[124]   Frescura FAM, Engelbrecht CA, Frank BS. Significance of periodogram peaks and a pulsation mode analysis of the Beta Cephei star V403 Car. Mon Not R Astron Soc 2008; 388: 1693-1707.
[125]   Dreschoff GAM, Zeller EJ. 415-year Greenland ice core record of solar proton events dated by volcanic eruptive episodes. In: Wakeffield D, Ed. TER-QUA Symposium Series 2. Lincoln, Nebraska: Nebraska Academy of Sciences, 1994; pp. 1-24.
[126]   Huang NE, Shen Z, Long SR, *et al*. The empirical mode decomposition and the Hilbert spectrum for nonlinear and nonstationary time series analysis. Proc Roy Soc London A 1998; 454: 903-995
[127]   Von Storch H, Zwiers FW. Statistical analysis in climate research. Cambridge: Springer/ Cambridge University Press 1999.
[128]   Levi Vehel J. Signal enhancement based on Holder regularity analysis. IMA volumes in mathematics and its application 2002; 132: 197-209.
[129]   Makarenko N, Karimova L, Kuandykov Y. Enhancement of the prediction of geophysical time series by modifying the regularity structure of a signal. Multi-wavelength investigation of solar activity. Proc. IAU Symp. № 223. Cambridge Univ Press 2004; pp. 707-708.
[130]   Traversi R, Usoskin IG, Solanki SK, *et al*. Nitrate in polar ice: a new tracer of solar variability. Solar Phys 2012; 280: 237-254.
[131]   Konstantinov BP, Kocharov GE. Astrophysical phenomena and radiocarbon. Dokl Akad Nauk SSSR 1965; 165: 63-64 (in Russian).
[132]   Kocharov GE, Zhorzholiani IV, Lomtatidze ZV, *et al*. Characteristics of solar activity during the last 400 years. Soviet Astron Lett 1990; 16: 723-728 (in Russian).
[133]   Stuiver M, Braziunas F. Atmospheric ^{14}C and century-scale solar oscillations. Nature 1989; 338: 405-407.
[134]   Vasiliev VA, Kocharov GE. On the solar activity dynamics during Maunder Minimum. In: Kocharov GE, Ed. Proc of XIII International Leningrad seminar on cosmophysics. Leningrad 1983; pp.75-101.
[135]   Dergachev V, Veksler V. Application of the radiocarbon method for studies of the environment in the past. USSR, Leningrad: Ioffe Phys-Tech Inst 1991 (in Russian).
[136]   Kovaltsov GA, Mishev A, Usoskin IG. A new model of cosmogenic production of radiocarbon 14C in the atmosphere, Earth Planet Sci Lett 2012; 337: 114-120.
[137]   O'Brien K. Secular variations in the production of cosmogenic isotopes in the Earth's atmosphere. J Geophys Res 1979; 84: 423-431.
[138]   Light ES, Merker M, Verschell HJ, *et al*. Time dependent worldwide distribution of atmospheric neutrons and of their products, 2, Calculation. J Geophys Res 1973; 78: 2741-2747.
[139]   Lingenfelter RE, Ramaty R. Radiocarbon variations and absolute chronology. In: Olsson IU, Ed. Proc of XII Nobel symp. Held in the Institute of physics, Upsala university. Stockholm: Almquist and Wiksell 1970; 513-535.
[140]   Kocharov GE, Akhmetkereev SKh, Peristykh AN. On the solar flare effect in the atmospheric radiocarbon. In: Kocharov GE, Ed. Possibilities of the methods of measurement of ultra-minor amount of isotopes. Leningrad: PhTI 1990; 45-70 (in Russian).

[141] Allan W, Lowe DC. A simple procedure for evaluating global cosmogenic 14C production in the atmosphere using neutron monitor data. Radiocarbon 2002; 44: 149-157

[142] Kocharov GE, Bitvinskas TT, Vasiliev VA, *et al.* Cosmogenic isotopes and astrophysical phenomena. In: Kocharov GE, Ed. Astrophysical phenomena and radiocarbon. Leningrad: PhTI 1985; 9-143 (in Russian).

[143] Korff SA, Mendell RB. Variations in radiocarbon production in the Earth's atmosphere. Radiocarbon 1980; 22: 159-165.

[144] Stuiver M, Quay PD. Changes in atmospheric Carbon-14 attributed to a variable sun Science 1980; 207: 11 -19.

[145] Kocharov GE, Koudriavtsev IV, Ogurtsov MG, *et al.* The nitrate content of Greenland ice and solar activity. Astr Rep 2000; 44: 825-829.

[146] Casatagnoli G, Lal D. Solar modulation effects in terrestrial production of carbon-14. Radiocarbon 1980; 22: 133-158.

[147] Dergachev VA, Kocharov GE, Metzkhvarishvili RYa, Zhorzholiani IV. Radioactive carbon in the Earth's atmosphere. In: VI Leningrad International seminar: "Acceleration of particles and nuclear processes in the space". Leningrad 1974; pp. 177-210 (in Russian).

[148] Stuiver M, Quay PD. Atmospheric ^{14}C changes resulting from fossil fuel CO_2 release and cosmic ray flux variability. Earth Planet Sci Lett 1981; 53: 349-362.

[149] Konstantinov AN, Kocharov GE, Levchenko VA. Cosmogenic isotopes ^{10}Be, ^{14}C, ^{36}Cl and astrophysical phenomena. In: Solar activity and solar-terrestrial relations. Leningrad: PhTI 1987; pp. 100-135.

[150] IPCC, Houghton *et al.*, Eds, Third assessment report. Climate change 2001: the scientific basis, Cambridge University Press 2001.

[151] Stuiver M, Grootes PM, Braziunas TF. The GISP2 18O climate record of the past 16,500 years and the role of the sun, ocean and volcanoes. Quat Res 1995; 44: 341-354.

[152] Stuiver M, Polach HA. Discussion: reporting of 14C data. Radiocarbon 1977; 19: 355-363.

[153] Southon J. Are the fractionation corrections correct: are the isotopic shifts for 14C/12C ratios in physical processes and chemical reactions really twice those for 13C/12C? Radiocarbon 2011; 3: 691-704.

[154] Stern MJ, Vogel PC. Relative 14C-13C kinetic isotope effects. J Chem Phys 1971; 55: 2007-2013.

[155] Reimer P J, Baillie MGL, Bard E, *et al.* IntCal09 and Marine09 radiocarbon age calibration curves, 0-50 000 years cal BP. Radiocarbon 2009; 51: 1111-1150.

[156] Heikkila U, Beer J, Feichter J. Modeling cosmogenic radionuclides ^{10}Be and ^{7}Be during the Maunder Minimum using the ECHAM5-HAM general circulation model. Atmos Chem Phys 2008; 8: 2797-2809.

[157] Field C, Schmidt G, Koch D, Salyk C. Modeling production and climate related impacts on ^{10}Be concentration in ice cores. J Geophys Res 2006: 111; D15107, doi:10.1029/2005JD00640.

[158] Raisbeck GM, Yiou F, Fruneau M, *et al.* Cosmogenic 10Be/7Be as a probe of atmospheric transport processes. Geophys Res Lett 1981; 8: 1015-1018.

[159] Steig EJ, Polissar PJ, Stuiver M, *et al.* Large amplitude solar modulation cycles of 10Be in Antarctica: implications for atmospheric mixing processes and interpretation of the ice core record. Geophys Res Let 1996; 23: 523-526.

[160] Heikkilä U, Beer J, Feichter J. Meridional transport and deposition of atmospheric ^{10}Be. Atmos Chem Phys 2009; 9: 515-527.

[161] Lal D. Theoretically expected variations in the terrestrial cosmic ray production rates of isotopes. In: Soc Italiana di Fisica-Bologna-Italy XCV corso 1988; pp. 216-233.

[162] Bard E, Raisbeck G, Yiou F, Jouzel J. Solar irradiance during the last 1200 years based on cosmogenic nuclides. Tellus B 2000; 52: 985-992.

[163] Lal D. ^{10}Be in polar ice: data reflect changes in cosmic ray flux or polar meteorology. Geophys Res Lett 1987; 14: 785-788.

[164] Svensson A, Andersen KK, Bigler M., *et al.* A 60 000 year Greenland stratigraphic ice core chronology, Clim. Past 2008: 4; 47-57.

[165] Vinther BM, Clausen HB, Johnsen SJ. *et al.* A synchronized dating of three Greenland ice cores throughout the Holocene. J Geophys Res 2006; 111: D13102 doi:10.1029/2005JD006921.

[166] Mc Corkell R, Fireman EL, Langway CC. Aluminium-26 and Beryllium-10 in Greenland Ice. Science 1967; 158: 1690-1695.

[167] Raisbeck GM, Yiou F, Fruneau M. *et al.* Measurements of ^{10}Be in 1000- and 5000-year-old Antarctic ice. Nature 1978; 275: 731-736.

[168] Kocharov GE, Vasiliev VA, Dergachev VA, Ostryakov VM. An 8000-year sequence of galactic cosmic ray fluctuations. Soviet Astron Lett 1983; 9: 110-112 (in Russian).

[169] Oeshger H, Siegenthaler U, Schotterer U, Gugelmann A. A box diffusiom model to study the carbon dioxide exchange in nature. Tellus 1974; 27: 168-192.

[170] Siegenthaler U, Heimann M, Oeshger H. 14C variations caused by changes in the global carbon cycle. Radiocarbon 1980; 22: 177-191.

[171] Stuiver M, Reimer PJ, Bard E. *et al.* INTCAL98 Radiocarbon age calibration, 24 000-0 cal BP. Radiocarbon 1998; 40: 1041-1083.

[172] Broecker WS. Geochemical tracers and ocean circulations. In: Warren BA, Wunsch C, Eds. Evolution of physical oceanography, scientific surveys in honor of Henry Stommel. Cambridge, Massachusetts: The MIT Press 1981; pp. 434-460.

[173] Dorman LI. Peculiarities of study of the cosmic ray variations by means of radiocarbon method. In: Proc. of the VI whole-union congress devoted to the problem "Astrophysical phenomena and radiocarbon". Tbilisi 1978; pp. 49-97 (in Russian).

[174] Kocharov GE, Tleugaliev SKh. Nine-box model of carbon-exchange processes. In: Kocharov GE, Ed. Possibilities of the methods of measurement of ultra-minor amount of isotopes. Leningrad: PhTI 1990; 76-94 (in Russian).

[175] Randerson T, Enting IG, Schuur EAG. *et al.* Seasonal and latitudinal variability of troposphere $\Delta^{14}CO_2$: post bomb contributions from fossil fuels, oceans, the stratosphere and the terrestrial biosphere. Global biogeochemical cycles 2002; 16: 1112, doi:10.1029/2002GB001876.

[176] Solanki S, Usoskin IG, Kromer B, *et al.* Unusual activity of the Sun during recent decades compared to the previous 11,000 years. Nature 2004; 431: 1084-1087.

[177] Damon PE, Sonett CP. Solar and terrestrial components of the atmospheric ^{14}C variation spectrum. In: Sonett CP, Giampapa MS, Mathews MS, Eds. The Sun in time. Tuscon, Univ. of Arizona press 1991; pp. 360-368.

[178] Stuiver M, Braziunas T. Sun, ocean, climate and atmospheric CO_2. The Holocene 1993; 3: 289-305.

[179] Bard E, Raisbeck GM, Yiou F, Jouzel J. Solar modulation of cosmogenic nuclide production over the last millennium: comparison between 14C and 10Be records. Earth and Planet Sci Lett 1997; 150: 453-462.

[180] Usoskin IG, Kromer B. Reconstruction of the cosmogenic 14C production rate from measured relative abundance. Radiocarbon 2000; 47: 31-37.

[181] Usoskin IG, Mursula K, Solanki S, *et al.* Reconstruction of solar activity for the last millennium using ^{10}Be data. Astron Astrophys 2004; 413: 745-751.

[182] Suess HE. Radiocarbon concentration in modern wood. Science 1955; 122: 415-417.

[183] Budyko MI, Izrael YA. Anthropogenic climate cange. University of Arizona Press: Tucson 1991.

[184] Keeling CD. Industrial production of carbon dioxide from fossil fuels and limestone. Tellus 1973; 25: 174-198

[185] Joos F. The anthropogenic perturbation of atmospheric CO2 and the climate system. In Bhattacharya SK, Mal TK, Chakrabarti S, Eds. Recent research developments in biotechnology and bioengineering. Special Issue: Biotechnology and Bioengineering of CO2 Fixation. Research Signpost 2003; pp. 1-21.

[186] Kocharov GE, Dergachev VA, Sementsov AA, *et al.* Concentration of radiocarbon in tree rings at 1564-1583, 1593-1615, 1688-1712. In: Kocharov GE, Dergachev VF, Eds. Proc. 5th all-union conference "Astrophysicaland radiocarbon". Tdilisi 1974; pp. 47-63. (in Russian)

[187] Damon PE, Eastoe CI, Mikheeva IB. The Maunder Minimum: an interlaboratory comparison of Δ^{14}C from AD 1688 to AD 1710. Radiocarbon 1999; 41: 47-51.

[188] Zhorzholiani IV, Kereselidze PG, Kocharov GE, *et al.* Measurement of radiocarbon content in tree rings for the period 1600-1940 with corrections for isotopic fractionation. In: Kocharov GE, Ed., Experimental Methods in Research of Astrophysical and Geophysical Phenomena. Leningrad: PhTI 1988; pp. 92-114. (in Russian)

[189] Masuda K, Furuzava H, Miyahara H *et al.* Radiocarbon content in Japanese Cedar during the Maunder minimum. In: Proc. of 28th Int. Cosmic Ray Conf 2003; pp. 4143-4146.

[190]   Kocharov GE, Ogurtsov MG, Tsereteli ZL. Cosmogenic radiocarbon as a means of studying solar activity in the past. Astron Rep 2003; 47: 1054-1062.

[191]   Vasiliev VA, Kocharov GE, Ogurtsov MG. Time-spectral analysis of the data on concentration of the cosmogenic isotopes in terrestrial archives. Izv RAN, Physics; 61: 1224-1227 (in Russian).

[192]   Damon PE, Burr G, Cain WJ, Donahue DJ. Anomalous 11-year $\Delta^{14}C$ cycle at high latitude? Radiocarbon 1992; 34: 235-239.

[193]   Mc Cormac FG, Reimer PJ, Hogg AG, *et al.* Calibration of the radiocarbon time scale for the Southern Hemisphere: AD 1850-950. Radiocarbon 2002; 44: 641-1651.

[194]   Fan CY, Chen TM, Yun SX, Dai KM. Radiocarbon activity variation in dated tree rings grown in McKenzie delta. Radiocarbon 1986; 28: 300-305.

[195]   Baxter MS, Farmer JG. Radiocarbon: short-term variations. Earth and Planet Sci Lett 1973; 20: 295-303.

[196]   Ogurtsov MG. On the possibility of forecasting the Sun's activity using radiocarbon solar proxy. Solar Phys 2005; 231: 167-176.

[197]   Nagovitsin YuA, Ivanov VG, Miletsky EV, *et al.* Solar activity reconstruction from proxy data. In.: Proc. of International conference-workshop "Cosmogenic climate forcing factors during the last millennium". Kaunas: Vytautas Magnus University 2003; pp. 41-50.

[198]   Ogurtsov MG. Solar paleoastrophysics as a tool for the future solar activity forecasting. In: Proc of the IX International conference on solar physics: solar activity as a factor of space weather. Pulkovo, St. Petersburg: GAO RAN 2005; pp. 209-214 (in Russian).

[199]   Korte M, Muscheler RJ. Centennial to millennial geomagnetic field variations. Space Weather Space Clim 2012; 2: A08.

[200]   Teanby N, Gubbins D. The effects of aliasing and lock-in processes on palaeosecular variation records from sediments. Geophys J Int 2000; 142: 563-570.

[201]   Kono M, Tanaka H. SEDI Symposium on reversals, secular variations and dynamo theory. Abstracts. Santa Fe (New Mexico) 1990; Sess. 5.

[202]   Mc Elhinny MW, Senanayake WE. Variations in the geomagnetic dipole, 1, the past 50000 years. J Geomagn Geoelectr 1982; 34: 39-51.

[203]   Yang S, Odah H, Shaw J. Variations in the geomagnetic dipole moment over the last 12,000 years. Geophyis J Int 2000; 140: 158-162.

[204]   Degachev VA, Dmitriev PB. Periodic fluctuations of radiocarbon abundance over the long time scale. In: Modern problems of solar cyclicity. St. Petersburg 1997; pp. 328-330.

[205]   Sonett PC. The present status of understanding of the long-period spectrum of radiocarbon. In: Taylor RE, Long A, Kra RS, Eds. Radiocarbon after four decades; an interdisciplinary perspective. New York: Springer-Verlag 1992; pp 50-61.

[206]   Voss H, Kurths J, Schwarz U. Reconstruction of grand minima of solar activity from 14C data: Linear and nonlinear signal analysis. J Geophys Res 1996; 101: 15 637-15 644.

[207]   Matsumoto K. Radiocarbon-based circulation age of the world oceans, J Geophys Res 2007; 112: C09004, doi:10.1029/2007JC004095.

[208]   Stuiver M, Quay PD, Ostlund HG. Abyssal water carbon-14 distribution and the age of the world oceans. Science 1983; 219: 849-851.

[209]   Garrison T. Oceanography: An invitation to marine science. Belmont, California: Brooks/Cole-Thomson Learning 2005; 522 pp.

[210]   Stommel H, Arons AB. On the abyssal circulation of the world ocean. I. Stationary planetary flow patterns on a sphere. Deep-Sea Res 1960; 6: 140-154.

[211]   Levitus S. Climatological atlas of the World Ocean. NOAA Prof. Rockville: Pap. 13 1982; 173 p.

[212]   Kunze E, Sanford TB. Abyssal mixing: where it is not. J Phys Oceanogr 1996; 26: 2286-2296.

[213]   Bryan K, Lewis L J. A water mass model of the world ocean. J Geophys Res 1979; 84: 2503-2517.

[214]   Tsujino H, Hasumi H, Suginohara N. Deep Pacific circulation controlled by vertical diffusivity at the lower thermocline depths. J Phys Oceanogr 2001; 30: 2853-2865

[215]   King FD, Deval AH. Estimates of vertical eddy diffusion through the thermohaline from phytoplankton nitrate uptake rates in the mixed layer of the eastern Pacific. Limnol Oceanogr 1979; 24: 645-651.

[216]   Stuiver M. ^{14}C distribution in the Atlantic ocean. J Geophys Res 1980; 85: 2711-2718.

[217] Ledwell JR, Watson AJ, Law CS. Evidence for slow mixing across the pycnocline from an open-ocean tracer-release experiment. *Nature* 1993; 364: 701-703.

[218] Schmitt RW, Ledwell JR, Montgomery ET. *et al.* Enhanced diapycnal mixing by salt fingers in the thermocline of the tropical Atlantic. Science 2005; 308: 685-688.

[219] Munk WH. Abyssal recipes. *Deep-Sea Res* 1966; 13: 707-730.

[220] Roemmich D, Hautala TS, Rudnick DL. Northward abyssal transport through the Samoan Passage and adjacent regions. *J Geophys Res* 1996; 101: 14039-14055.

[221] Ledwell JR, Montgomery ET, Polzin KL, *et al.* Evidence for enhanced mixing over rough topography in the abyssal ocean. Nature 2000; 403: 179-182.

[222] Beer J, Andree M, Oeschger M, Stauffer B. Temporal [10]Be variation in ice. Radiocarbon 1983; 2: 269-278.

[223] Fligge M, Solanki SK, Beer J. Determination of solar cycle length variations using the continuous wavelet transform. Astr Astrophys 1999; 346: 313-321.

[224] Steig J, Morse DL, Waddington ED, Polissar PJ. Using the sunspot cycle to date ice cores. Geophysical Res. Letters 1998; 25: 163-166.

[225] Beer J, Tobias S, Weiss N. An active Sun throughout the Maunder minimum. Solar Physics 1998; 181: 237-249.

[226] Usoskin IG, Mursula K, Kovaltsov GA. Heliospheric modulation of cosmic rays and solar activity during the Maunder minimum. J Geophys Res 2001; 106: 16039-16046.

[227] Muscheler R, Beer J. Solar forced Dansgaard/Oeschger events?, Geophys Res Lett 2006; 33: L20706, doi:10.1029/2006GL026779.

[228] Ogurtsov MG. Long_term solar cycles according to data on the cosmogenic beryllium concentration in ice of Central Greenland. Geomagn Aeron 2010; 50: 475-481.

[229] Alley RB, Finkel RC, Nishiizumi K, *et al.* Changes in continental and sea-salt atmospheric loadings in central Greenland during the most recent deglaciation: Model-based estimates. J Glaciol 1995; 41: 503-514.

[230] Finkel RC, Nishiizumi K. Beryllium 10 concentrations in the Greenland Ice Sheet Project 2 ice core from 3-40 ka. J Geophys Res 1997; 102: 26699-26706.

[231] Mayewski PA, Meeker LD, Twickler MS, *et al.* Major features and forcing of high-latitude northern hemisphere atmospheric circulation using a 110,000-year-long glaciochemical series. J Geophys Res 1997; 102: 26345-26366.

[232] Yang Q, Mayewski PA, Whitlow SI, *et al.* Global perspective of nitrate flux in ice cores. J Geophys Res 1995; 100: 5113-5121.

[233] Guyodo Y, Valet J.-P. Global changes in geomagnetic intensity during the past 800 thousand years. Nature 1999; 399: 249-252.

[234] Knudsen MF, Riisager P, Donadini F, *et al.* Variations in the geomagnetic dipole moment during the Holocene and the past 50 kyr. Earth Planet Sci Lett 2008; 272: 319-329.

[235] Bond G, Kromer B, Beer J, *et al.* Persistent Solar Influence on North Atlantic Climate during the Holocene 2001; Science 294: 2130 -2136.

[236] Dreschoff GAM, Zeller EJ. Ultra-high Resolution Nitrate in Polar Ice as Indicator of Past Solar Activity. Solar Phys 1998; 177: 365-374.

[237] Legrand MR, Stordal F, Isaksen ISA, Rognerud B. A model study of the stratospheric budget of odd nitrogen, including effects of solar cycle variations. Tellus 1989; 41B: 413-426.

[238] Legrand MR, Kirchner S. Origins and variation of nitrate in South Polar precipitations. J Geoph Res 1990; 95: 3493-3507.

[239] Mayewski PA, Meeker LD, Morrison MC, *et al.* Greenland Ice Core «Signal» Characteristics: an Expanded View of Climate Change. J Geophys Res 1993; 98: 12839-12847.

[240] Zeller EJ, Parker BC. Nitrate ions in Antarctic firn as a marker of solar activity. Geophys Res Lett 1981; 8: 895-898.

[241] Logan JA. Nitrogen oxides in the troposphere: global and regional budgets. J Geophys Res 1983; 88: 10785-10805.

[242] Mayewski PA, Lyons WB, Twickler MS, *et al.* An ice record of atmospheric response to antropohenic sulphate and nitrate. Nature 1990; 346: 554-556.

[243] Dreschhoff GAM, Zeller EJ. Evidence of individual solar proton events in Antarctic snow. Solar Phys 1990; 127: 333-346.

[244] Jackman CH, DeLand MT, Labow GJ, *et al*. Satellite measurements of middle atmospheric impacts by solar proton events in solar cycle 23. Space Sci. Rev 2006; 125: 381-391.

[245] Jackman HC, Fleming LE, Vitt FM. Influence of extremely large solar proton events in a changing atmosphere. J Geophys Res 2000; 105: 11659-11670.

[246] Reagan JB, Meyerott RE, Nightingale RW, *et al*. Effects of the August 1972 solar particle events on stratospheric ozone, J Geophys Res 1981; 86: 1473-1494

[247] Müller R, Crutzen PJ. A possible role of galactic cosmic rays in chlorine activation during polar night. J Geophys Res 1993; 98: 20483-20490.

[248] Webber W. The production of free elements in the ionospheric D-layer by solar and galactic cosmic rays and the resultant absorption of radiowaves. J Geophys Res 1962; 67: 5091-5106.

[249] Dunkerton T. On the mean meridional mass motions of the stratosphere and mesosphere. J Atmosph Physics 1978; 35: 2325-2332.

[250] Zeller EJ, Dreshhoff GAM, Laird CM. Nitrate flux on the Ross Ice Shelf, Antarctica, and its relation to solar cosmic rays. Geophys Res Letters 1986; 13: 264-1267.

[251] Röthlisberger R, Hutterli MA, Wolff E, *et al*. 2002. Nitrate in Greenland and Antarctic ice cores: a detailed description of post-depositional processes. Annals Glaciol; 35: 209-216.

[252] Randall C E, Harvey VL, Manney GL, *et al*. Stratospheric effects of energetic particle precipitation in 2003-2004. Geophys Res Lett 2005; 32: L05802, doi:10.1029/2004GL022003.

[253] Hoch SW, Schelander P, Bourgeois S, *et al*. Energy balance at summit, Greenland, 2001-2002. EGU04-J-04254 2004.

[254] Carroll JJ. On the determinations of the near surface temperature regime of the south Polar plateau. *J Geophys Res* 1984; 89: 4941-4952.

[255] Kotlyakov VM, Krenke AN. Glaciers as climate indicators. Izv AN SSSR 1982; 18: 1215-1228 (in Russian).

[256] Dreschhoff GAM, Laird CM. Evidence for a stratigraphic record of supernovae in polar ice. Adv Space Res 2006; 38: 1307-1311.

[257] Kepko L, Spence H, Smart DF, Shea MA. Interhemispheric observations of impulsive nitrate enhancements associated with the four large ground-level solar cosmic ray events (1940-1950). J. Atmosph Sol.-Terr Phys. 200; 971: 1840-1845.

[258] Mc Cracken KG, Dreschhoff GAM, Smart DF, Shea MA. Solar cosmic ray events for the period 1561-1994: 2. The Gleissberg periodicity. J Geophys Res 2001A; 106: 21599-21610.

[259] Palmer AS, van Ommen TD, Curran AJ, Morgan V. Ice-Core evidence for a small-source of atmospheric nitrate. J Geophys Res 2001; 28: 1953-1956.

[260] Mc Cracken KG, Dreschhoff GAM, Zeller EJ, *et al*. Solar cosmic ray events for the period 1561-1994. 1. Identification in Polar Ice, 1561-1950. J Geophys Res 2001B; 106: 21585-21598.

[261] Usoskin IG, Solanki S, Kovaltsov GA, *et al*. Solar proton events in cosmogenic isotope data. Geophys Res Lett 2006; 33: L08107

[262] Kostantinov AN, Levchenko VA, Kocharov GE, *et al*. Theoretical and experimental aspects of solar flares manifestation in radiocarbon abundance in tree rings. Radiocarbon 1992; 34: 247-253.

[263] Wolff, E. W., Bigler M., Curran M. A. J., *et al*. The Carrington event not observed in most ice core nitrate records. Geophys Res Lett 2012 ; 39 : L08503.

[264] Ogurtsov MG., Jungner H, Kocharov GE, *et al*., Nitrate concentration in Greenland ice: an indicator of changes in fluxes of solar and galactic high-energy particles. Solar Phys 2004; 222: 177-190

[265] Wang N, Yao T, Thompson L. Concentration of nitrate in the Guliya ice core from the Qinghai-Xizang Plateau and the solar activity. Chinese Sci Bull 1998; 43: 841-844.

[266] Wittmann AD. The sunspot cycle before the Maunder Minimum. Astron Astrophys 1978; 66: 93-97.

[267] Yau KKC, Stephenson FR. A revised catalogue of Far Eastern observations of sunspots (165 BC to AD 1918). Quart J Roy Astron Soc 1988; 29: 175-197.

[268] Willis DM, Davda VN, Stephenson FR. Comparison between Oriental and Occidental sunspot observations. Quart J Roy Astron Soc 1996; 37: 189-229.

[269] Nagovitsyn YuA. A nonlinear mathematical model for the solar cyclicity and prospects for reconstructing the solar activity in the past. Astron Lett 1997; 23: 742 - 748.

[270] Ogurtsov MG. Cosmogenic isotopes and their role in present-day solar paleoastrophysics. Geomagn Aeron 2007; 47: 85-93.

[271] Usoskin IG, Solanki SK, Kovaltsov GA. Grand minima and maxima of solar activity: new observational constraints. Astron Astrophys 2007; 471: 301-309.

[272] Muscheler R, Joos F, Muller SA, *et al*. Climate: How unusual is today's solar activity? Nature 2005; 436: E3-E4.

[273] Ogurtsov MG. Was the solar activity in the last 100 years abnormally high? On the quality of modern solar paleoreconstructions. Astron Lett 2007B; 33: 419-426.

[274] Beer J, Vonmoos MV, Muscheler R *et al*. Heliospheric Modulation over the past 10,000 Years as derived from Cosmogenic Nuclides. In: Kajita T, Asaoka Y, Kawachi A, Matsubara Y, Sasaki M, Eds. 28th International Cosmic Ray Conference, July 31 - August 7 2003, Universal Academy Press, Tokyo, Japan 2003; 7: pp. 4147-4150.

[275] Damon PE, Jirikowic JL. Solar forcing of global climate change? In: Talor RE, Long A, Kra RS, Eds. Radiocarbon after four decades: an interdisciplinary perspective. New York, Springer-Verlag 1992, pp 117-129.

[276] Vasiliev SS, Dergachev VA. The ~2400-year cycle in atmospheric radiocarbon concentration: bispectrum of ^{14}C data over the last 8000 years. Annales Geophysicae 2002; 20: 115-120.

[277] Schmidt B, Gruhle W. Radiokohlenstoffgehalt und dendrochronologie. Naturwissenshaftliche Rundschau 1988; 5: 177-182.

[278] Usoskin IG. A history of solar activity over millennia. Living Rev Solar Phys 2013; 10: 1.

[279] Reedy RC. Recent solar energetic particles updates and trends. In: Lunar and Planetary Science XXXIII, Abstracts of papers submitted to the conference, March 11 - 15, 2002. Lunar nd Planetary Institute, Houston, U.S.A 2002; 33: 1938-1940.

[280] Reedy RC. Constraints on solar particle events from comparisons of recent events and million-year averages. Astron Soc Pacific Conference Series 1996; 95: 429-436.

[281] Hoyt HP, Walker RM, Zimmerman DW. Solar flare proton spectrum averaged over the last 5×10^3 years. In: Proc Lunar Sci Conf. 4th 1973; pp. 2489-2502.

[282] Boekl RS. A depth profile of ^{14}C in lunar rock 12002. Earth Planet Sci Letters 1972; 16: 269-272.

[283] Fink D, Klein J, Middleton R, *et al*. ^{41}Ca, ^{26}Al, and ^{10}Be in lunar basalt 74275 and ^{10}Be in the double drive tube 74002/74001. Geochim Cosmochim Acta 1998; 62: 2389-2402.

[284] Kohl CP, Murrell MT, Russ GP, Arnold JR. Evidence for the constancy of the solar cosmic ray flux over the past 10 million years: ^{53}Mn and ^{26}Al measurements. In: Proc Lunar Sci Conf 9th 1978; pp. 2299-2310.

[285] Usoskin IG, Kromer B, Ludlow F, *et al*. The AD775 cosmic event revisited: the Sun is to blame. Astron Astrophys 2013; 552: L3.

[286] Bonino G, Castagnoli GC, Bhandari N, Taricco C. Behavior of the Heliosphere over prolonged solar quiet periods by 44Ti measurements in meteorites. Science 1995; 270: 1648-1650.

[287] Heymann M. The evolution of climate ideas and knowledge. WIREs Clim Change 2010; 1: 581-597, doi: 10.1002/wcc.6.

[288] Shvedov PN. Tree as archive of droughts. Meteorological bulletin 1892, No.5 (in Russian).

[289] Dergachev VA. Radiocarbon chronometer. Priroda 1994; 1: 3-15 (in Russian).

[290] Douglass AE. A method of estimatig rainfall by the growth of trees. In: Huntington E., Ed. The climatic factor, Carnegie Institute Washington Publications 1914; pp. 101-122.

[291] Evans MN, Reichert BK, Kaplan A, *et al*. A forward modeling approach to paleoclimatic interpretation of tree-ring data. J Geophys Res 2006; 111: G03008, doi:10.1029/2006JG000166.

[292] Gates DM. Biophysical Ecology. New York: Springer 1980; 611 pp.

[293] Gou X, Zhou F, Zhang Y, *et al*. Forward modeling analysis of regional scale tree-ring patterns around the northeastern Tibetan Plateau, Northwest China. Biogeosciences Discuss 2013; 10: 9969-9988.

[294] Fritts H. Tree rings and climate. Acad Press: London, 567 P.

[295] Cook ER, Kairiukstis LA Methods of Dendrochronology: Applications in the Environmental Sciences. Kluwer Academic Publishers: Dordrecht 1989.

[296] Briffa KR. Annual climate variability in the Holocene: interpreting the message of ancient trees. Quat Sci Rev 2000; 19; 87-105.

[297] Briffa KR, Osborn TJ, Schweingruber FH, Harris IC, Jones PD, Shiyatov SG, Vaganov EA. Low frequency temperature variations from a northern tree-ring density network. Journal of Geophysical Research 2001;106 2929-2941.

[298]   Esper J, Cook ER, Schweingruber FH. Low-frequency signals in long tree-ring chronologies for reconstructing past temperature variability. Science 2002; 295: 2250-2253.

[299]   Jones PD, Briffa KR, Barnett TP, Tett SFB. High-Resolution palaeoclimatic records for the last millennium: interpretation, integration and comparison with general circulation model control-run temperatures. The Holocene 1998; 8.4: 455-471.

[300]   Mann M, Hughes M. Tree ring chronologies and climate variability. Science 2002; 296: 848-852.

[301]   Mann MN, Zhang Z, Hughes MK. *et al*. Proxy-Based reconstructions of hemispheric and global surface temperature variations over the past two millennia. Proc Nat Acad Sci 2008; 105: 13252-13257.

[302]   Jalkanen R, Tuovinen M. Annual needle production and height growth: better climate predictors than radial growth at treeline? Dendrochronologia 2001; 19: 39-44.

[303]   Pensa M, Sepp M, Jalkanen R. Connections between climatic variables and the growth and needle dynamics of Scots Pine (Pinus sylvestris L.) in Estonia and Lapland. Int J Biometeorology 2006; 50: 205-214.

[304]   Lindholm M, Aalto T, Grudd H, *et al.*: Common temperature signal in four well-replicated tree growth series from northern Fennoscandia. J Quat Sci 2012; 27: 828-834.

[305]   Helama S, Holopainen J, Timonen M, *et al*. Comparison of living-tree and subfossil ring-widths with summer temperatures from 18th, 19th and 20th centuries in Northern Finland. Dendrochronologia 2004; 21/3: 147-154.

[306]   Lindholm M, Eronen M. A reconstruction of mid-summer temperatures from ring-widths of Scots pine since AD 50 in northern Fennoscandia. Geografiska Annaler 2000; 82: 527-535.

[307]   Helama S, Seppa H, Birks HJB, Bjune AE. Reconciling pollen-stratigraphical and tree-ring evidence for high- and low-frequency temperature variability in the past millennium. Quat Sci Rev 2010; 29: 3905-3918.

[308]   Helama S, Seppa H, Bjune AE, Birks HJB. Fusing pollen-stratigraphic and dendroclimatic proxy data to reconstruct summer temperature variability during the past 7.5 ka in subarctic Fennoscandia. J Paleolimnol 2012; 48:275-286, DOI 10.1007/s10933-012-9598-1.

[309]   Briffa KR, Jones PD, Bartholin TS, *et al*. Fennoscandian summers from A.D. 500: temperature changes on short and long timescales. Clim Dyn 1992; 7: 111-119.

[310]   Briffa KR, Jones PD, Schweingruber FH, *et al*. Unusual twentieth-century summer warmth in a 1,000-year temperature record from Siberia. Nature 1995; 376: 156-159.

[311]   Cook ER, Buckley RD, D'Arrigo RD, Peterson MJ. Warm-season temperatures since 1600 BC reconstructed from Tasmanian tree rings and their relationship to large-scale sea surface temperature anomalies 2000. Clim Dyn; 16: 79-91.

[312]   Briffa KR, Jones PD, Schweingruber FH, Osborn TJ. Influence of volcanic eruptions on Northern Hemisphere summer temperature over the past 600 Years. Nature 1998; 393: 450-455.

[313]   D'Arrigo R, Wilson R, Liepert B, Cherubini P. On the 'divergence problem' in northern forests: A review of the tree-ring evidence and possible causes. Global Planet Change 2007; doi:10.1016/j.gloplacha.2007.03.004.

[314]   Esper J, Frank D. Divergence pitfalls in tree-ring research. Climatic Change 2009; 94:261-266.

[315]   Jones PD, Parker DE, Osborn TJ, Briffa KR. Global and hemispheric temperature anomalies - land and marine records. In: Trends: A compendium of data on global change. Carbon dioxide information analysis center, Oak Ridge National Laboratory, US Department of Energy, Oak Ridge, Tenn., USA 2001.

[316]   Lloyd A, Fastie C. Spatial and temporal variability in the growth and climate response of treeline trees in Alaska. Clim Change 2002; 58: 481-509.

[317]   Vaganov E, Hughes M, Kirdyanov A, *et al*. Influence of snowfall and melt timing on tree growth in Subarctic Eurasia. Nature 1999; 400: 149-151.

[318]   Liepert B. Observed reductions in solar surface radiation in the United States and worldwide from 1961 to 1990. Geophys Res Lett 2002; 29: 1421, 10.1029/2002GL014910.

[319]   Stanhill G, Cohen S. Global dimming: a review of the evidence for a widespread and significant reduction in global radiation with discussion of its probable causes and possible agricultural consequences. Agricultural Forest Meteorology 2001; 107: 255-278.

[320]   Loehle C. A Mathematical Analysis of the Divergence Problem in Dendroclimatology. Clim Change 2008; DOI 10.1007/s10584-008-9488-8.

[321]   Urey HC. Oxygen isotopes in nature and in the laboratory. Science1948; 108: 602-603.

[322]   Dansgaard W. The O18-abundance in Fresh Water. Geochimica et Cosmochimica Acta 1954; 6: 241-260.

[323]   Tian L, Yao T, Schuster PF, *et al.* Oxygen-18 concentrations in recent precipitation and ice cores on the Tibetan Plateau. J Geophys Res 2003; 108: 4293.

[324]   Morgan V, van Ommen TD. Seasonality in late-Holocene climate from ice core records. The Holocene 1997; 7: 351-354.

[325]   Vasil'chuk YuK, Kotlyakov VM. Principles of isotope geocryology and glaciology. Moscow: Moscow University Press 2000; (in Russian).

[326]   Farquhar GD, O'Leary MH, Berry JA. On the relationship between carbon isotope discrimination and the intercellular carbon dioxide concentration in leaves. Australian Journal of Plant Physiology 1982; 9: 121-137.

[327]   Hilasvuori E, Berninger F, Sonninen E, *et al.* Stability of climate signal in carbon and oxygen isotope records and ring width from Scots Pine (*Pinus sylvestris* L.) in Finland. J Quat Sci 2009; 24: 469-480.

[328]   Seftigen K, Linderholm HW, Loader NJ, *et al.* The influence of climate on 13C/12C and 18O/16O ratios in tree ring cellulose of Pinus sylvestris L. growing in the central Scandinavian Mountains. Chemical Geology 2011; 286: 84-93.

[329]   Grinstead MJ, Wilson AT, Ferguson CW. ^{13}C/^{12}C ratio variations in Pinus Longaeva (Bristole Pine) cellulose during the last millennium. Earth Plane Sci Lett 1979; 42: 251-253.

[330]   Tans PP, Mook WG. Past atmospheric CO_2 levels and the ^{13}C/^{12}C ratios in tree rings. Tellus 1980; 32b: 268-283.

[331]   Gagen M, Zorita E, McCarroll D, *et al.* Cloud response to summer temperatures in Fennoscandia over the last thousand years. Geophys Res Lett 2011; 38: L05701.

[332]   Lipp J, Trimborn P, Fritz P. Stable isotopes in tree ring cellulose and climatic change. Tellus 1991; 43B: 322-330.

[333]   McCarroll D, Tuovinen M, Campbell R, *et al.* A critical evaluation of multi-proxy dendroclimatology in northern Finland. J Quat Sci 2011; 26: 7-14.

[334]   Robertson I, Switsur VR, Carter AHC, *et al.*, Signal strength and climate relationship in ^{13}C/^{12}C ratios of tree ring cellulose from oak in east England. J Geophys Res 1997B; 102: 19507-19516.

[335]   Saurer M, Siegenthaler U, Schweingruber F. The climate-carbon isotope relationship in tree rings and the significance of site conditions. Tellus 1995; 47B: 320-330.

[336]   Beltrami H. Climate from Borehole Data: Energy Fluxes and Temperatures Since 1500. Geophys Res Lett 2002; 29: 2111, doi:10.1029/2002GL015702.

[337]   Mann ME, Rutherford S, Bradley RS, *et al.* Optimal surface temperature reconstructions using terrestrial borehole data. J Geophysl Res 2003; 108: 4203, doi:10.1029/2002JD002532.

[338]   Rutherford S, Mann ME. Correction to "Optimal surface temperature reconstructions using terrestrial borehole data". J Geophys Res 2004; 109: D11107, doi:10.1029/2003JD004290.

[339]   IPCC, Solomon, S.; Qin, D.; Manning, M.; Chen, Z.; Marquis, M.; Avery, K.B.; Tignor, M.; and Miller, H.L., Ed., Climate Change 2007: The Physical Science Basis, Contribution of Working Group I to the Fourth Assessment Report of the Intergovernmental Panel on Climate Change.Cambridge University Press 2007.

[340]   Basha MM. Altawfegat Al elhameah, Cairo 1890.

[341]   Basha AS. Tagweem Alnial, Cairo 1916

[342]   Borisenkov EP. Development of fuel and energy base and its influence on weather and climate. Meteorology and Hydrology 1977; 2: 3-14 (in Russian).

[343]   Dobrovolny P, Moberg A, Brazdil R. *et al.* Monthly, seasonal and annual temperature reconstructions for Central Europe derived from documentary evidence and instrumental records since AD 1500. Climatic Change 2010; 101: 69-107.

[344]   Krenke AN. Mass-exchange in glacier systems over the USSR territory. Leningrad: Hydrometeoizdat 1982; 288 p.

[345]   Tarussov A. The Arctic from Svalbard to Severnaya Zemlya: climatic reconstructions from ice cores. In: Bradley RS, Jones PD, Eds. Climate Since A.D. 1500. Routledge 1992; pp. 505-516

[346]   Ekaykin AA, Lipenkov VYa, Barkov NI, *et al.* Snow accumulation change over the last 350 years at the slope of the Antarctic ice sheet 200 km inland from Mirny Station. Kriosphera Zemli 2000; 4: 57-66. (in Russian).

[347]   Cuffey KM, Paterson WSB. *The Physics of Glaciers, Fourth Edition,* Elsevier 2010; 693 pp.

[348]   Barnett TP, Santer BD, Jones PD *et al.* Estimates of low frequency natural variability in near-surface air temperature. The Holocene 1996; 6.3: 225-263.

[349]   Briffa KR, Osborn TJ. Blowing hot and cold. Science 2002; 295: 2227-2228.

[350]   Luckman BH, Briffa KR, Jones PD, Schweingruber FH. Summer temperatures at the Columbia Icefield, Alberta, Canada, 1073-1987. The Holocene 1997; 7: 375-389.

[351]   Briffa KR, Jones PD, Schweingruber FH. Tree-ring density reconstructions of summer temperature patterns across western North America since 1600. J Climate 1992; 5: 735-754.

[352]   D'Arrigo RD, Jacoby GC Jr. Dendroclimatic evidence from northern North America. In: Bradley RS, Jones PD, Eds. Climate science AD 1500 Routledge: London 1992; pp. 296-311.

[353]   Lara A, Villalba R. A 3620-year temperature reconstruction from Fiztroya Curpessoids tree rings in southern South America. Science 1993; 260: 1104-1106.

[354]   Parker DE, Legg TP, Folland CK. A new daily Central England temperature series. Int. J Climatol 1992; 12: 317-342.

[355]   Pfister C. Monthly temperature and precipitation in central Europe 1525-1979: quantifying documentary evidence on weather and its effects. In: Bradley RS, Jones PD, Eds. Climate science AD 1500. Routledge: London 1992; pp. 549-571.

[356]   Fisher DA, Koerner RM, Kuivinen K. *et al.* Intercomparison of ice-core (Δ ^{18}O) and precipitation records from sites in Canada and Greenland over the last 3500 years and over the last few centuries in detail using EOF techniques. In: Jones PD, Bradley SR, Jozel J, *eds.* Climatic variations and forcing mechanisms of the last 2000 years. Berlin: Springer 1996; pp. 297-328.

[357]   Dunbar RB, Wellington GM, Colgan M, Glynn PW. Eastern Pacific sea surface temperature 1600 AD: the d ^{18}O record of climate variability in Galapagos corals. Paleoceanography 1994; 9: 291-315.

[358]   Quinn TM, Crowley TJ, Taylor FW, *et al.* Multicentury stable isotope record from a New Caledonia coral: interannual and decadal sea surface temperature wariability in the Southwest Pacific since 1657 A.D. Paleoceanography 1998; 13: 412-426.

[359]   Mann ME, Bradley RS, Hughes MK. Northern Hemisphere temperatures during the past millennium: inferences, uncertainties, and limitations. Geophys Res Lett 1999; 26: 759-762.

[360]   Mc Carroll D, Jalkanen R, Hicks S, *et al.* Multiproxy dendroclimatology: a pilot study in northern Finland. The Holocene 2003; 13: 829-838.

[361]   McCarroll D, Loader NJ, Jalkanen R, *et al.*, A 1200-year multi-proxy record of tree growth and summer temperature at the northern pine forest limit of Europe. The Holocene 2013; 23: 471-484.

[362]   Crowley TJ, Lowery TS. How warm was the Medieval warm period? Ambio 2000; 29; 51-54.

[363]   Moberg A, Sonechkin DM, Holmgren K, *et al.* High variable Northern Hemisphere temperatures reconstructed from low- and high-resolution proxy data. Nature 2005; 433: 613-617.

[364]   Loehle CA. 2000-year global temperature reconstruction based on non-treering proxies. Energy Envir 2007; 18: 1049-1058.

[365]   Ogurtsov MG, Jungner H, Helama S, *et al.* Paleoclimatological evidence for unprecedented recent temperature rise at the extratropical part of the Northern Hemisphere. Geografiska Annaler 2011; 93: 17-27.

[366]   Milankovitch M. Theorie mathematique des phenomenes thermiques produits par la radiation solaire. Zagreb, Paris : Acad Yougoslave Sci Arts 1920; 633 p.

[367]   *Monin* AS, *Shishkov* YA. Climate as a problem of physics. *Physics-Uspekhi* 2000; 43: 381-406.

[368]   Petit JR, Jouzel J, Raynaud D, *et al.* Climate and atmospheric history of the past 420,000 years from the Vostok Ice Core, Antarctica. Nature 1999; 399: 429-436.

[369]   Berger A, Loutre MF. Insolation values for the climate of the last 10 million years. Quat Sci Rev 1991; 10; 297-317.

[370]   Lisiecki LE, Raymo ME. A Pliocene-Pleistocene stack of 57 globally distributed benthic Δ^{18}O records. Paleoceanography, 2005; 20: PA1003, doi:10.1029/2004PA001071.

[371]   Shackleton NJ. The 100,000-year ice-age cycle identified and found to lag temperature, carbon dioxide, and orbital eccentricity. Science 2000; 289: 1897-1902.

[372]   Berger AL, Loutre MF. An exceptionally long interglacial ahead? Science 2002; 297: 1287-1288.

[373]   Loutre MF, Berger AL. Future climate changes: Are we entering an exceptionally long interglacial? Clim Change 2000; 46: 61-90.

[374]   Johnsen SJ, Dansgaard W, Clausen HB, Langway CC. Oxygen isotope profiles through the Antarctic and Greenland ice sheets. Nature 1972; 235; 429 - 434.

[375]   Masson-Delmotte V., *et al.* Rapid climate variability during warm and cold periods in polar regions and Europe. Comptes Rendus Geoscience 2005; 337: 935-946.

[376]   Grootes PM, Stuiver M. Oxygen 18/16 variability in Greenland snow and ice with 10^3 to 10^5-year time resolution. J Geophys Res 1997; 102: 26455-26470.

[377]   Stuiver M, Grootes P. Trees and the ancient record of heliomagnetic cosmic ray flux modulation. In: Pepin RO, Eddy JA, Merrill RB, Ed. Proc Conf Ancient Sun. New York, Oxford, Toronto, Sydney: Pergamon Press 1995; pp. 165-175.

[378]   Dansgaard W. Evidence for general instability of past climate from a 250-kyr icecore record. Nature 1993; 364: 218-220.

[379]   Johnsen SJ, Clausen HB, Dansgaard W, *et al.* Irregular glacial interstadials recorded in a new Greenland ice core. Nature 1992; 359: 311- 313.

[380]   Landais A, Masson-Delmotte V, Jouzel J *et al.* The glacial inception as recorded in the North GRIP Greenland ice core: timing, structure and associated abrupt temperature changes. Clim Dyn 2006; 26: 273-284.

[381]   Mann ME, Park J, Bradley RS. Global interdecadal and century-scale climate oscillations during the past five centuries. Nature 1995; 378: 266-270.

[382]   Mahasenan N, Watts RG, Dowlatabady H. Low-frequency oscillations in temperature-proxy records and implications for recent climate change. Geophys Res Lett 1997; 24: 563-566.

[383]   Shabalova MV, Weber SL. Patterns of temperature variability on multidecadal to centennial timescales. J Geophys Res 1999; 104: 31023-31042.

[384]   Ogurtsov MG, Jungner H, Lindholm M. A potential century-scale rhythm in six major paleoclimatic records in the Northern Hemisphere. Geografiska Annaler 2007; 89: 129-136.

[385]   Ogurtsov M, Lindholm M, Jalkanen R. Global warming- scientific facts, problems and possible scenarios. In: Tarhule A, Ed. Climate variability- regional and thematic patterns. Rieka: InTech 2013; pp. 75-103.

[386]   Ogurtsov MG, Lindholm M. Uncertainties in assessing global warming during the 20th century: disagreement between key data sources. Energy Envir 2006;17: 685-706.

[387]   Bürger G. Clustering climate reconstructions. Climate of the Past Discussions 2010; 6: 659-679.

[388]   Lamb HH. The early medieval warm epoch and its sequel. Palaeogeogr Palaeoclimatol Palaeoecol 1965; 1: 13-37.

[389]   Hughes MK, Diaz HF. Was there a Medieval Warm Period, and if so, where and when? *Clim. Change* 1994; 26: 109-142.

[390]   Ljungqvist FC. A new reconstruction of temperature variability in the extra-tropical Northern Hemisphere during the last two millennia. Geografiska Annaler 2010; 92: 339-351.

[391]   Mann ME, Zhang Z, Rutherford S, *et al.* Global signatures and dynamical origins of the Little Ice Age and Medieval Climate anomaly. Science 2009; 326: 1256-1260.

[392]   Mc Intyre S, McKitrick R. Corrections to the Mann *et al.* (1998) proxy data base and Northern Hemisphere average temperature series. Energy Envir 2003;14: 751-772.

[393]   Mc Intyre S, McKitrick R. Hockey sticks, principal components and spurious significance. Geophys Res Lett 2005a; 32: L03710, doi:2004GL021750.

[394]   Mc Intyre S, McKitrick R. Reply to Comment by von Storch and Zorita on "Hockey Sticks, Principal Components and Spurious Significance". Geophys Res Lett 2005b; 32: L20714, doi:10.1029/2005GL023089.

[395]   Mc Intyre S, McKitrick R. Reply to Comment by Huybers on "Hockey Sticks, Principal Components and Spurious Significance". Geophysl Res Lett 2005c; 32: L20714, doi:10.1029/2005GL023586.

[396]   Mann ME, Bradley RS, Hughes MK. Global-scale temperature patterns and climate forcing over the past six centuries. Nature 1998; 392: 779-787.

[397]   Huybers P. Comment on ''Hockey Sticks, Principal Components, and Spurious Significance'' by S. McIntyre and R. McKitrick. Geophys Res Lett 2005; 32: L20705, doi:10.1029/2005GL023395

[398]   von Storch H, Zorita E, Jones PD, *et al.* Reconstructing past climate from noisy data. Science 2004; 306: 679-682.

[399]   von Storch H, Zorita E. Comment on "Hockey Sticks, Principal Components, and Spurious Significance" by S. McIntyre and R. McKitrick. Geophys Res Lett 2005; 32: L20701, doi:10.1029/2005GL022753.

[400]   Wahl ER, Ammann CM. Robustness of the Mann, Bradley, Hughes reconstruction of Northern Hemisphere surface temperatures: Examination of criticisms based on the nature and processing of proxy climate evidence. Clim Change 2007; 85: 33-69, doi:10.1007/s10584-006-9105-7.

[401]   IPCC, Stocker TF, Qin D, Plattner G-K, *et al.*, Eds. Summary for Policymakers. In: Climate Change 2013: The Physical Science Basis. Contribution of Working Group I to the Fifth Assessment Report of the Intergovernmental Panel on Climate Change. Cambridge, United Kingdom and New York: Cambridge University Press 2013.

[402]   Wilson R, D'Arrigo R, Buckley B, *et al*. A matter of divergence—tracking recent warming at hemispheric scales using tree-ring data. J Geophys Res 2007; 112: D17103.1-D17103.17.

[403]   Rind D. The Sun's role in climate variations. Science 2002; 296: 673-677.

[404]   Henderson-Sellers A, Robinson PJ. Contemporary climatology. New York: London Scientific and Technical, John Wiley and Sons Inc. 1986; 439 p.

[405]   Sellers WD. Physical climatology. Chicago and London: the University of Chicago Press 1965; 272 p.

[406]   Shipper TR, Clough SA, Brown PD, *et al*. Spectral Cloud Emissivities from LBLRTM/AERI QME. In: Proc of the Eighth Atmospheric Radiation Measurement (ARM) Science Team Meeting 1998; pp. 688-92.

[407]   Kirkby J. Cloud: A particle beam facility to investigate the effect of cosmic rays on clouds. In: Kirkby J, Ed. Proc. of the workshop on ion-aerosol-cloud interactions. Geneva: CERN 2001-007 2001; pp. 175-248.

[408]   Wang K, Wan Z, Wang P, *et al*. Estimation of surface long wave radiation and broadband emissivity using Moderate Resolution Imaging Spectroradiometer (MODIS) land surface temperature/emissivity products. J Geophys Res 2005; 110: D11109, doi:10.1029/ 2004JD005566.

[409]   Pudovkin MI. Effect of the solar activity on the lower atmosphere and weather. Soros Educational Journal 1996; 10; 106-114 (in Russian).

[410]   Fröhlich C. Solar irradiance variability since 1978: revision of the PMOD composite during solar cycle 21. Space Sci Rev 2006; 125: 53-65.

[411]   Dewitte S, Crommelynck D, Mekaoui S, Joukoff A. Measurement and uncertainty of the long-term total solar irradiance trend. Solar Phys 2004; 224: 209-216.

[412]   Mordvinov AV. Reduction of measurements of total solar flux to a unified scale and uncertainty in estimating its long-term trends. Geomagn Aeron 2010; 50: 933-936.

[413]   Kopp G, Lean JL. A new, lower value of total solar irradiance: Evidence and climate significance. Geophys Res Lett 2011; 38: L01706, doi:10.1029/2010GL045777.

[414]   Haigh JD. Mechanisms for solar influence on the Earth's climate. In: Tsuda T, Fujii R, Shibata K, Geller MA, Eds. Climate and weather of the Sun-Earth system (CAWSES): selected papers from the 2007 Kyoto Symposium. Tokyo: Terrapub 2009; pp. 231-256.

[415]   Waple AM, Mann ME, Bradley RS. Long-term patterns of solar irradiance forcing in model experiments and proxy based surface temperature reconstructions. Clim Dynam 2002; 18: 563-578.

[416]   Reid G. Solar total irradiance variations and the global sea surface temperature record. J Geophys Res 1991; 96: 2835-2844.

[417]   Schuurmans CJE, Oort AH. A statistical study of pressure changes in the troposphere and lower stratosphere after strong solar flares. Pure Appl Geophys 1969; 75: 233-246.

[418]   Dmitriev AA, Lomakina TYu. Cloudiness and X-radiation from the space. In: Rakipova LR, Ed. Effects of solar activity in the lower atmosphere. Leningrad 1977; pp. 70-77 (in Russian).

[419]   Kondratyev KYa, Nikolsky GA. The stratospheric mechanism of solar and anthropogenic influences on climate. In: McCormac EM, Seliga TA, Eds. Solar-Terrestr Influences on Weather and Climate. Holland: D. Reidel Publ Corp 1979; pp. 311-322.

[420]   Kondratyev KYa, Nikolsky GA. The solar constant and climate. Sol Phys 1983; 89: 215-222.

[421]   Pudovkin MI, Babushkina SV. Atmospheric transparency variations associated with geomagnetic disturbances. J Atmos Solar-Terr Phys. 1992A; 54: 1135-1138.

[422]   Veretenenko SV, Pudovkin MI. Effects of cosmic ray variations in the circulation of the lower atmosphere. Geomagn Aeron 1993; 33: 35-40.

[423]   Blinova EN. Hydrodynamic theory of pressure waves, temperature waves and centers of action of the atmosphere. Dokl Acad Nauk SSSR 1943; 39: 284-287 (in Russian).

[424]   Pudovkin MI. Effect of the solar activity on the lower atmosphere and weather. Soros Educational Journal 1996; 10; 106-114 (in Russian).

[425]   Pudovkin MI, Babushkina SV. Effect of electromagnetic and corpuscular radiation of solar flares on the intensity of zonal atmospheric circulation. Geomagn Aeron 1991; 31: 388-392.

[426]   Pudovkin MI, Babushkina SV. Influence of solar flares and disturbances of the interplanetary medium on the atmospheric circulation. J Atmosph Solar-Terr Phys 1992B; 54: 841-846.

[427]   Roldugin VK, Vashenyk EV. Atmospheric transparency variations caused by cosmic rays. Geomagn Aeron 1994; 34: 251-253.

[428]   Veretenenko SV, Pudovkin MI. The galactic cosmic ray Forbush decrease effects on total cloudiness variations. Geomagn Aeron 1994; 34: 463-468.

[429]   Todd M, Kniveton DR. Short-term variability in satellite-derived cloud cover and galactic cosmic rays: an update. J of Atmosph and Solar-Terr Phys 2004; 66: 1205-1211.

[430]   Starkov GV, Roldugin VK. Relationship between atmospheric transparency variations and geomagnetic activity. Geomagn Aeron 1995; 34: 559-562.

[431]   Veretenenko SV, Pudovkin MI. Variations of solar radiation input to the lower atmosphere associated with different helio/geophysical factors. J Atmosph Solar-Terr Phys 1999; 61: 521-529.

[432]   Veretenenko SV, Pudovkin MI. Global radiation changes in the lower atmosphere related to solar activity phenomena. Int J Geomagn Aeron 2001; 3: 1-7.

[433]   Vanhellemont F, Fussen D, Bingen C. A time series analysis to investigate a possible link between cosmic rays and stratospheric aerosols. Adv Space Res 2003; 31: 2145-2150.

[434]   Harrison RG, Stephenson DB. Empirical evidence for a nonlinear effect of galactic cosmic rays on clouds. Proc R Soc A 2006; 462: 1221-1233.

[435]   Kancirova M, Kudela K. Cloud cover and cosmic ray variations at Lomnicky stit high altitude observing site. Atmos Res 2014, 149:166-173.

[436]   Labitzke K, van Loon H. Associations between the 11-year solar cycle, the QBO and the atmosphere, Part I: The troposphere and stratosphere in the northern hemisphere in winter, J Atmos Terr Phys 1988; 50: 197-206.

[437]   Mustel ER. The modern state of the problem of the corpuscular-atmospheric link reality. In: Solar-atmospheric relationship in the theory of climate and weather forecasts. Leningrad: Hydrometeoizdat 1974, pp. 7 -18 (in Russian).

[438]   Roldugin VC, Tinsley BA. Atmospheric transparency changes associated with solar J Atmosph Solar-Terr Phys 2004; 66: 1143-1149.

[439]   Roldugin VK, Starkov GV. Dependence of atmospheric transparency variation on solar activity. Studia Geophysica and Geodaetica 1998; 42: 137-146.

[440]   Shumilov OI, Vashenyuk EV, Kasatkina EA, Baidanov SA. Increase of stratospheric aerosols after solar proton events. Atmospheric ozone. Proc SPIE 1993; 2047: 70-82.

[441]   Shumilov OI, Kasatkina EA, Henriksen K, Vashenyuk EV. Enhancement of stratospheric aerosol after solar proton event. Annales Geophys 1996; 4: 1119-1123.

[442]   Marichev VN, Bogdanov VV, Zhivet'ev IV, Shvetsov BM. Influence of the geomagnetic disturbances on the formation of aerosol layers in the stratosphere. Geomagn Aeron 2004; 44: 779-786.

[443]   Mironova IA, Pudovkin MI. Increase in the aerosol content of the lower atmosphere after the solar proton flares in January and August 2002 according to data of lidar observations in Europe. *Geomagn Aeron* 2005; 45: 221-226.

[444]   Goncharenko YV, Kivva FV. On the sizes of atmospheric aerosol particles in the reflecting layers appearing after strong solar flares. Radiophysics and Electronics 2002; 7: 509-512 (in Russian).

[445]   Gerding M, Baumgarten G, Blum U. *et al.* Observation of unusual mid-stratospheric aerosol layer in the Arctic: possible sources and implications for polar vortex dynamics. Annales Geophys. 2003; 21: 1057-1069.

[446]   Kasatkina EA, Shumilov OI. Cosmic ray-induced stratospheric aerosols: A possible connection to polar ozone depletions. Annales Geophysicae 2005; 23: 675-679.

[447]   Vanhellemont F, Fussen D, Bingen C. Cosmic rays and stratospheric aerosols: Evidence for a connection? Geophys Res Lett 2002; 29: 1715, 10.1029/2002GL015567.

[448]   Bingen C, Vanhellemont F, Fussen D. A new regularized inversion method for the retrieval of stratospheric aerosol size distributions applied to 16 years of SAGE II data (1984-2000): method, results and validation. Annales Geophysicae 2003; 21: 797-804.

[449]   Veretenenko SV, Ivlev LS, Uliev VA. Research of stratospheric aerosol variations during solar proton events of January 2005. Probl Arctic Antarctic 2008; 3: 126-130 (in Russian).

[450]   Mironova IA, Usoskin IG, Kovaltsov GA, Petelina SV. Possible effect of extreme solar energetic particle event of 20 January 2005 on polar stratospheric aerosols: direct observational evidence. Atmos Chem Phys 2012; 12: 769-778.

[451]   Mironova IA, Usoskin IG. Possible effect of extreme solar energetic particle events of September-October 1989 on polar stratospheric aerosols: a case study. Atmos Chem Phys 2013; 13: 8543-8550

[452]   Svensmark H, Bondo T, Svensmark J. Cosmic ray decreases affect atmospheric aerosols and clouds. Geophys Res Let 2009; 36: L15101, doi:10.1029/2009GL038429.

[453]   Calogovic J, Albert C, Arnold F, *et al.* Sudden cosmic ray decreases: No change of global cloud cover. Geophys Res Lett 2010; 37: L03802, doi:10.1029/2009GL041327.

[454]   Svensmark H, Friis-Christensen E. Variation of cosmic ray flux and global cloud coverage - a missing link in solar-climate relationships. J Atmosph Solar-Terr Phys 1997; 59: 1225-1232.

[455]   Kernthaler SC, Toumi R., Haigh JD. Some doubts concerning a link between cosmic ray fluxes and global cloudiness. Geophys Res Lett 1999; 26(7): 863-865.

[456]   Jorgensen TS, Hansen AW. Comments on "Variation of cosmic ray flux and global cloud coverage - a missing link in solar-climate relationships" by Henrik Svensmark and Eigil Friis-Christensen. J Atmos Solar-Terr Phys 2000; 62: 73-77.

[457]   Kristjansson JE, Kristiansen J. Is there a cosmic ray signal in recent variations in global cloudiness and cloud radiative forcing? J Geophys Res 2000; 105: 11851-11863.

[458]   Svensmark H, Friis-Christensen E. Reply to comments on "Variation of cosmic ray flux and global cloud coverage: A missing link in solar-climate relationships". J Atmosph Solar-Terrestr Phys 2000; 62: 79- 80.

[459]   Marsh N, Svensmark H. Galactic Cosmic ray and El Nin~o-Southern Oscillation trends in ISCCP-D2 low-cloud properties. J Geophys Res 2003; 108: 4195, doi:10.1029/2001JD001264.

[460]   Palle E, Butler CJ. The influence of cosmic rays on terrestrial clouds and global warming. Astronomy and Geophys 2000; 41: 18-22.

[461]   Kristjansson JE, Staple A, Kristiansen J, Kaas E. A new look at possible connections between solar activity, clouds and climate. Geophys Res Lett 2002; 29: 2107-2111.

[462]   Sloan T, Wolfendale AW. Testing the proposed causal link between cosmic rays and cloud cover. Environm Res Lett 2008; 3: 024001.

[463]   Erlykin AD, Wolfendale AW. Cosmic Ray Effects on cloud cover and their relevance to climate change. J Atmos Solar-Terr Phys 2011;73: 1681-1686.

[464]   Evan AT, Heidinger AK, Vimon DJ. Arguments against a physical long-term trend in global ISCCP cloud amounts. Geophys Res Lett 2007; 34: L04701, doi:10.1029/2006GL028083.

[465]   Gray LJ, Beer J, Geller M, *et al.* Solar influences on climate. Reviews in Geophysics 2010; 48: RG4001, doi:10.1029/2009RG000282.

[466]   Laken BA, Palle´ ED, Calogovic J, Dunne EM. A cosmic ray-climate link and cloud observations. J Space Weather Space Clim 2012; 2: A18.

[467]   Voiculescu M, Usoskin I. Persistent solar signatures in cloud cover: spatial and temporal analysis. Environ Res Lett 2012; 7: 044004.

[468]   Artamonova IV, Veretenenko SV. Galactic cosmic ray variation influence on baric system dynamics at middle latitudes. J. Atmosph Sol-Terr Phys 2011; 73: 366-370.

[469]   Veretenenko S, Thejll P. Influence of energetic Solar Proton Events on the development of cyclonic processes at extratropical latitudes. IOP Publishing. J Phys: Conf Ser 2013; 409: 012237, doi:10.1088/1742-6596/409/1/012237.

[470]   Veretenenko SV, Thejll P. Solar proton events and evolution of cyclones in the North Atlantic. Geomagn Aeronomy 2008; 48: 542 -552.

[471]   Artamonova IV, Veretenenko SV. Atmosphere pressure variations at high latitudes associated with Forbush decreases of cosmic rays. In: Proceedings of the 9th Intl Conf. "Problems of Geocosmos". St. Petersburg, Russia 2012; pp. 188-192.

[472] Veretenenko S, Ogurtsov M. Regional and temporal variability of solar activity and galactic cosmic ray effects on the lower atmosphere circulation. Adv Space Res 2012. 49: 770-783.

[473] Vangengeim GYa. Backgrounds of the microcirculation method of meteorological predictions for the Arctic. Tr Arct Nauch-Issled Inst 1952; 34: 314-322 (in Russian)

[474] Girs A.A. Macrocirculation method of long-term meteorological predictions. Leningrad: Hydrometeoizdat 1974 (in Russian).

[475] AARI. Department of long-term meteorological forecasts of Arctic and Antarctic Research institute, private communication; 2013.

[476] Khromov SP, Petrosyants MA. Meteorology and Climatology. Moscow: Moscow State Univ. 1994 (in Russian).

[477] Maliniemi V, Asikainen T, Mursula K, Seppala A. QBO-dependent relation between electron precipitation and wintertime surface temperature, J Geophys Res Atmos 2013; 118, doi:10.1002/jgrd.50518.

[478] Tinsley BA, Deen GW. Apparent tropospheric response to MeV-GeV particle flux variations: a connection *via* electrofreezing of supercooled water in high-level clouds. J Geophys Res 1991; 96: 22283-22296.

[479] Laken BA, Kniveton DR, Frogley MR. Cosmic Rays Linked to Rapid Mid-Latitude Cloud Changes. Atmosph Phys Discussions 2010; 10: 18235-18253.

[480] Svensmark J, Enghoff MB, Svensmark H. Effects of cosmic ray decreases on cloud microphysics. Atmos Chem Phys Discuss 2012; 12: 3595-3617.

[481] Pudovkin MI, Raspopov OM. Mechanism of the influence of solar activity on the state of lower atmosphere and meteoparameters. Geomagn Aeron 1992; 32: 1-22 (in Russian).

[482] Raspopov OM, Shumilov OI, Kasatkina EA. Cosmic rays as a primary factor of effect of solar variability on climatic and atmospheric parameters. Biophysics 1998; 43: 902-908 (in Russian).

[483] Tinsley BA. Influence of solar wind on the global electric circuit, and inferred effects on cloud microphysics, temperature, and dynamics in the troposphere. Space Sci Rev 2000; 94: 231-258.

[484] Tinsley BA. The global atmospheric electric circuit and its effects on cloud microphysics. Rep Prog Phys 2008; 71: 066801.

[485] Harrison RG, Ambaum MHP. Observed atmospheric electricity effect on clouds. Environmental Res Lett 2009; 4: 014003.

[486] Stozhkov YI, Ermakov VI, Pokrevsky PE Cosmic rays and atmospheric processes Izv RAN, Ser Fiz 2001; 65: 406-10 (in Russian)

[487] De Jager C, Usoskin I. On possible drivers of Sun-induced climate changes. J Atmosph Solar-Terr Phys 2006; 68: 2053-2060.

[488] Raspopov OM, Lovelius NV, Shumilov OI, Kasatkina EA. The nonlinear character of the effect of solar activity on climatic processes. Geomagn Aeron 2001; 41: 407-412.

[489] White WB, Dettinger MD, Cayan DR. Global average upper ocean temperature response to changing solar irradiance: exciting the internal decadal mode. In: Proc. of 1st Solar and Space Weather Euroconference. The Solar cycle and Terrestrial Climate, Santa Cruze de Tenerife, Tenerife, Spain. ESA SP-463 2000; pp. 125-133.

[490] Zielinski G.A., Mayewski P.A., Meeker L.D. *et al.*: 1994. Record of volcanism since 7000 BC from the GISP2 Greenland ice core and implications for the volcano-climate system. Science. V. 264, P. 948-951.

[491] Briffa K, Osborn T, Schweingruber F. Large-scale temperature inferences from tree rings: a review. Global Planetary Change 2004; 40: 11-26.

[492] Bach W. Global air pollution and climatic change. Rev Geophys Space Phys 1976; 14: 429-434.

[493] Turco RP, Whitten RC, Toon OB, *et al*. OCS, stratospheric aerosols and climate. Nature 1980; 283: 283-286.

[494] Chernavskaya MM, Cherenkova EA. Study of influence of the volcanic eruptions on the circulation processes over extratropical latitudes of the Northern hemisphere. Electr J Investigated in Russia 2004; 141: 1546-1556, http://zhurnal.ape.relarn.ru/ articles/2004/141.pdf.

[495] Pyle DM, Beattie PD, Bluth GJS. Sulfur emission to the stratosphere from explosive volcanic eruptions. Bull Volcanol 1996; 57: 663-671.

[496]   Eisel FL, Tanner DJ. Measurement of the gas phase concentration of H_2SO_4 and methane sulfuric acid and estimates of H_2SO_4 production and loss in the atmosphere. J Geophys Res 1993; 98(D5): 9001-9011.

[497]   Zielinski GA. Use of paleo-records in determining variability within the volcanism-climate system. Quat Sci Rev 2000; 19: 417-438.

[498]   Kondratyev KYa, Cracknell AP. Observing Global Climate Change. London: Taylor and Francis Ltd 1998; 562 pp.

[499]   Luther FM. Effect of increased stratospheric aerosol on planetary albedo. Preprint, Lawrence Livermore Lab, Univ Calif 1974.

[500]   Toon OB, Pollack JB. A global average model of atmospheric aerosol for radiative transfer calculations. J Appl Meteorol 1976; 15: 225-246.

[501]   Kondratyev KYa, Pozdnyakov DV. Aerosol models of the atmosphere. Moscow: Nauka 1981 (in Russian).

[502]   Stothers RB. The great Tambora eruption and its aftermath. Science 1984; 224: 1191-1198.

[503]   Sato M, Hansen JE, McCormick MP, Pollack JB. Stratospheric aerosol optical depths, 1850-1990. Journal Geophys Res 1993; 88: 22987-22994.

[504]   Charlson RJ, Schwartz SE., Hales JM, *et al.* Climate forcing by antropogenic aerosols. Science 1992; 255: 423-430.

[505]   Emile-Geay J, Seager R, Cane MA, *et al.* Volcanoes and ENSO over the past millennium. J Climate 2008; 21: 3134-3148.

[506]   Myhre G, Myhre A, Stordal F. Historical evolution of radiative forcing of climate. Atmos Environ 2001; 35: 2361-2373.

[507]   Andronova NG, Rozanov EV, Yang F, *et al.*: Radiative forcing by volcanic aerosols from 1850 to 1994. J Geophys Res 1999; 104, 16 807-16 821.

[508]   Rampino MR., Self S. Historic eruptions of Tambora (1815), Krakatau (1883) and Agung (1963), their stratospheric aerosol and climate impact. Quaternary Res 1982; 18: 127-143.

[509]   Tahira M, Nomura M, Sawada Y, Kamo K. Infrasonic and acoustic-gravity waves generated by the mount Pinatubo eruption of June 15, 1991. In: Newhall CG, Punongbayan RS, Eds. Fire and mud, eruptions and lahars of Mount Pinatubo, Phillippines. PHIVOLCS and Univ Wash Press 1997, pp. 601-613.

[510]   Harris BM, Highwood EJ. A simple relationship between volcanic sulfate aerosol optical depth and surface temperature change simulated in an atmosphere-ocean general circulation model. J Geophys Res 2011; 116: D05109, doi:10.1029/2010JD014581.

[511]   Sigurdsson H. Evidence of volcanic loading of the atmosphere and climate response. Global Planet Change1990; 3: 277-289.

[512]   Shumilov OI, Kasatkina EA, Raspopov OM, *et al.* An estimation of the climatic response to the variations in solar and volcanic activity. Geomagn Aeron 2000; 40: 687-691.

[513]   Zielinski GA, Mayewski PA, Meeker LD, *et al.* Potential atmospheric impact of the Toba mega-eruption ~71 000 years ago. Geophys. Res. Lett 1996; 23: 837-840.

[514]   Huang CY, Zhao M, Wang CC, Wei G. Cooling of the South-China Sea by the Toba eruption and correlation with other climate proxies ~71 000 years ago. Geophys Res Lett 2001; 28: 3915-3918.

[515]   Solanki SK, Krivova NA, Schlussler M, Fligge M. Search for a relationship between solar cycle amplitude and length. Astron and Astrophys 2002; 396: 1029-1035.

[516]   Damon PA, Peristykh AN. Solar cycle length and 20th century Northern Hemisphere warming. Geophys Res Lett 1999; 26: 2469-247

[517]   Laut P, Gundermann J. Solar cycle lengths and climate: a reference visited. J Geophys Res 2000; 105: 27489-27492.

[518]   Butler CJ. Maximum and minimum temperatures at Armagh Observatory, 1844-1992, and the length of the sunspot cycle. Solar Phys 1994; 152: 35-42.

[519]   Zhou K, Butler CJ. A statistical study of the relationship between the solar cycle length and tree-ring index values. Journ Atmosph Solar-Terr Phys 1998; 60: 1711-1718.

[520]   Ogurtsov MG. Secular variation in aerosol transparency of the atmosphere as the possible link between long-term variations in solar activity and climate. Geomagn Aeron 2007; 47: 118-128.

[521]   Bryson RA, Goodman BM. Volcanic activity and climatic changes. Science 1980; 207: 1041-1044.

[522]   Kondratyev KYa. Climatic effects of aerosol and clouds. Springer/Praxis UK: Chichester 1999.

[523] Hamill P, D'Auria R, Turco RP. An empirical approach to the nucleation of sulfuric acid droplets in the atmosphere. Annals of Geophysics 2003; 46: 331-340.

[524] Arnold F. Ion nucleation - a potential source for stratospheric aerosols. Nature1982; 299: 134-137.

[525] Arnold F, Buhrke Th. New H_2SO_4 and HNO_3 vapour measurements in the stratosphere - evidence for a volcanic influence. Nature1983; 301: 293-295.

[526] Hofmann DJ, Rosen JM. On the temporal variation of stratospheric aerosol size and mass during the first 18 months following the 1982 eruptions of El Chichon. Journal of Geophys Res 1984; 89: 4883-4890.

[527] Yu F. Binary H_2SO_4-H_2O homogeneous nucleation based on kinetic quasi-unary nucleation model: Look-up tables. J Geophys Res 2006; 111: D04201, doi:10.1029/2005JD006358

[528] Kazil J, Lovejoi ER. Tropospheric ionization and aerosol production: a model study. J Geophy Res 2004; 109: D19206.

[529] Marsh N, Svensmark H. Low cloud properties influenced by cosmic rays. Phys Rev Lett 2000; 85: 5004-5007.

[530] Neher HV. Cosmic-Rays at high latitudes and altitudes covering four solar maxima. J Geophys Res; 76: 1637-1651

[531] Bazilevskaya GA, Usoskin IG, Fluckiger EO *et al.* Cosmic ray induced ion production in the atmosphere. Space Sci Rev 2008; DOI 10.1007/s11214-008-9339-y.

[532] Wissing JM, Kallenrode M-B. Atmospheric Ionization Module Osnabru.ck (AIMOS): A 3-D model to determine atmospheric ionization by energetic charged particles from different populations. J Geophys Res 2009; 114, A06104, doi:10.1029/2008JA013884.

[533] . Yu F. Altitude variations of cosmic ray induced production of aerosols: Implications for global cloudiness and climate. J Geophys Res 2002; 107: 1118, doi:10.1029/2001JA000248.

[534] Hofmann DJ, Rosen JM. Condensation nuclei events at 30 km and possible influences of solar cosmic rays. Nature 1983; 302: 511-514.

[535] Harrison RG, Aplin KL. Atmospheric condensation nuclei formation and high-energy radiation. J Atmos Solar-Terr Phys 2001; 63: 1811-1819.

[536] Eichkorn S, Wilhelm S, Aufmhoff H. *et al.* Cosmic ray-induced aerosol-formation: First observational evidence from aircraft-based ion mass spectrometer measurements in the upper troposphere. Geophys Res Lett 2002; 29: 1698, DOI: 10.1029/2002GL015044.

[537] Svensmark H, Pedersen JO,.Marsh NM. *et al.* Experimental evidence for the role of ions in particle nucleation under atmospheric conditions. Proc Roy Soc A 2007; 463: 385-396.

[538] Enghoff MAB, Pedersen JOP, Bondo T., *et al.* Evidence for the role of ions in aerosol nucleation. J PhysChem A 2008; 112: 10305-10309.

[539] Kirkby J, Curtius J, Almeida J, *et al.* Role of sulphuric acid, ammonia and galactic cosmic rays in atmospheric aerosol nucleation. Nature 2011; 476: 429-433.

[540] Zuev VE, Kabanov MV. Optics of atmospheric aerosol. Lningrad: Hydrometeoizdat 1987 (in Russian).

[541] Hamill P, Toon OB. Microphysical processes affecting stratospheric aerosol particles. J Atmos Sci 1977; 34: 1104- 1118.

[542] Fuchs NA, Sutugin AG. High dispersed aerosols. In: Hidy GM, Brock JR, Eds. Topics in current aerosol research. Pergamon Press 1971; 2: pp. 1-60.

[543] Yu F. Formation of large NAT particles and denitrification in polar stratosphere: possible role of cosmic rays and effect of solar activity. Atmos Chem Phys 2004; 4: 2273-2283.

[544] Tolbert MA. Sulfate aerosols and polar stratospheric cloud formation. Science 1994; 264: 527-528.

[545] Kent GS, Poole LR, Mc Cormick MP. Characteristics of Arctic polar stratospheric clouds as measured by airborne lidar. J Atmos. Solar.-Terr. Phys. 1986; 43: 2149-2161.

[546] Pawson S, Naujokat B, Labitzke K. On the polar stratospheric cloud formation potential of the northern stratosphere. J Geophys Res 1995; 100: 23215-23226.

[547] Mohnen VA. Stratospheric ion and aerosol chemistry and possible links with cirrus cloud microphysics - a critical assessment. J Atmos Sci 1990; 47: 1933-1948.

[548] Avakyan SV, Voronin NA. Radio-optical and optical mechanisms of the influence of space factors on global climate warming. J Optical Technology (OSA). 2010; 77: 150-152.

[549] Koudriavtsev IV, Jungner H. Variations in atmospheric transparency under the action of galactic cosmic rays as a possible cause of their effect on the formation of cloudiness. 2011; 51: 656-663.

[550]   Haigh ID. The impact of solar variability on climate. Science 1996; 272: 981-984.

[551]   Shindell D, Rind D, Balachandran N, *et al*. Solar cycle variability, ozone and climate. Science 1999; 284: 305-308.

[552]   NIPCC-IR1 Idso CD, Carter MR, Singer SF, Eds. Climate Change Reconsidered: 2011 Interim Report of the Nongovernmental Panel on Climate Change (NIPCC), Chicago, IL: The Heartland Institute 2011.

[553]   Brohan P, Kennedy JJ, Haris I, *et al*. Uncertainty estimates in regional and global observed temperature changes: a new dataset from 1850. J Geophys Res 2006; 111: D12106, doi:10.1029/2005JD006548,

[554]   Mears CA, Wentz FJ. The effect of diurnal correction on satellite-derived lower troposphere temperature. Science 2005; 309: 1548-1551.

[555]   Christy JR, Norris WB. What may we conclude about global tropospheric temperature trends? Geophys Res Lett 2004; 31: L06621, doi:10.1029/ 2003GL019361.

[556]   Christy JR, Spencer RW, Mears CA, Wentz F. Correcting temperature data sets. Science 2005; 310: 972-973.

[557]   Huang SP, Pollack HN, Shen PY. Late Quaternary climate reconstruction based on borehole heat flux data, borehole temperature data and the instrumental record. Geophys Res Lett 2008; 35: L13703.

[558]   Douglass DH, Christy JR, Pearson BD, Singer SF. A comparison of tropical temperature trends with model predictions. Int J Climatology 2008; 28: 1693-1701.

[559]   Santer BD, Thorne PW, Haimberger L, *et al*. Consistency of modeled and observed temperature trends in the tropical troposphere. Int J Climatology 2008; 28: 1703-1722.

[560]   Fu Q, Manabe S, Johanson CM. On the warming in the tropical upper troposphere: models *versus* observations. Geophys Res Lett 2011; 38: L15704, doi:10.1029/2011GL048101.

[561]   Budyko MI. The effect of solar radiation variations on the climate of the Earth. Tellus 1969; 21: 611-619.

[562]   Sellers WD. A global climatic model based on the energy balance of the earth-atmosphere system. J Appl Meteorol 1969; 8: 392-400.

[563]   Manabe S, Wetherland RT. Thermal equilibrium of the atmosphere with a given distribution of relative humidity. J Atmosph Sci 1967; 24: 241-259.

[564]   Hoyt DV, Schatten KH. A discussion on plausible solar irradiance variations, 1700-1993. J Geophysl Res 1993; 98; 18895-18906.

[565]   Lean J, Beer J, Bradley R. Reconstruction of Solar Irradiance since 1610: Implications for Climate Change. Geophys Res Lett 1995; 22: 3195-3198.

[566]   Beer J, Mende W, Stellmacher R. The Role of the Sun in Climate Forcing. Quaternary Sci Rev 2000;19: 403-415.

[567]   Mordvinov AV, Makarenko NG, Ogurtsov MG, Jungner H. Reconstruction of magnetic activity of the Sun and changes in its irradiance on a millennium timescale using neurocomputing. Sol Phys 2004; 224: 247-253.

[568]   Shapiro AI, Schmutz W, Rozanov E, *et al*. A new approach to the long-term reconstruction of the solar irradiance leads to large historical solar forcing. Astron Astrophys 2011; 529: A67.

[569]   Lindzen RS, Giannitsis C. On the climatic implications of volcanic cooling. J Geophys Res 1998; 103: 5929-5941.

[570]   Lindzen RS, Choi Y-S. On the Determination of Climate Feedbacks from ERBE Data. Geophys Res Lett 2009; 36: L16705, doi:10.1029/2009GL039628.

[571]   Lindzen RS, Chou M-D, Hou AY. Does the Earth have an adaptive infrared iris? Bull American Meteorol Soc 2001; 82: 417-432.

[572]   Douglass DH, Knox RS, Pearson BD, Clark A Jr. Thermocline flux exchange during the Pinatubo event Geophys Res Lett 2006; 33: L19711, doi:10.1029/2006GL026355.

[573]   Schwartz SE. Reply to comments by G. Foster *et al*., R. Knutti *et al*., and N. Scafetta on "Heat capacity, time constant, and sensitivity of Earth's climate system". J Geophys Res 2008; 113: D15105, doi:10.1029/2008JD009872.

[574]   Chylek P, Lohmann U. Aerosol radiative forcing and climate sensitivity deduced from the last glacial maximum to Holocene transition. Geophys Res Lett 2008; 35: L23703, doi:10.1029/2008GL033888.

[575]   Chylek P, Lohmann U, Dubey M, *et al*. Limits on climate sensitivity derived from recent satellite and surface observations. J Geophys Res 2007; 112: D24S04, doi:10.1029/2007JD008740

[576]   Forster PM, Gregory JM. The climate sensitivity and its components diagnosed from Earth radiation budget data. J of Climate 2006; 19: 39-52.

[577]   Andronova, NG, Schlesinger ME. Causes of Global Temperature Changes During the 19th and 20th Centuries. Geophys Res Let 2000; 27: 2137-2140.

[578]   Knutti R, Allen MR., Friedlingstein P, *et al.* A review of uncertainties in global temperature projections over the twenty-first century. J Climate 2008; 21: 2651-2663.

[579]   Ogurtsov MG. On the possible contribution of solar-cosmic factors to a global warming of 20th century. Izv RAN, Physics 2007; 71: 1051-1053 (in Russian).

[580]   Mursula K, Usoskin IG, Kovaltsov GA. Reconstructing the long-term cosmic ray intensity: linear relations do not work. Annales Geophysicae 2003; 21: 863-867.

[581]   Stivens B, Bony S. What are climate models missing? Science 2013; 340: 1053-1054.

[582]   Solanki SK, Krivova NA. Can solar variability explain global warming since 1970? J Geophys Res 2003; 108: 1200-1212.

[583]   Benestad RE, Schmidt GA. Solar Trends and Global Warming. J Geophys Res 2009; 114: D14101, doi:10.1029/2008JD011639.

[584]   Scafetta N. Empirical evidence for a celestial origin of the climate oscillations and its implications. J Atmosph Solar-Terr Phys 2010; 72: 951-970.

[585]   Scafetta N. Multi-scale harmonic model for solar and climate cyclical variation throughout the Holocene based on Jupiter-Saturn tidal frequencies plus the 11-year solar dynamo cycle. J Atmosph Solar-Terr Phys 2012; 72: 951-970.

[586]   Palle E, Goode PR, Montanes-Rodriguez P and Koonin SE. Can Earth's albedo and surface temperatures increase together? Eos 2006; 70: 37-43.

[587]   Pinker RT, Zhang B, Dutton EG. Do satellites detect trends in surface solar radiation? *Science* 2005; 308: 850- 854.

[588]   Palle E, Goode PR, Montanes-Rodriguez P, Koonin SE. Changes in the Earth's reflectance over the past two decades; Science 2004: 1299-1301.

[589]   Wild M, Gilgen H, Roesch A, *et al.* From dimming to brightening: trends in solar radiation inferred from surface observations. Science 2005; 308: 847-850.

[590]   Sugihara G, May RM. Nonlinear forecasting as a way of distinguishing chaos from measurement error in time series. Nature 1990; 344: 734-741.

[591]   Legasov VA, Kuzmin II, Chernoplekov AN. Izv AN SSSR 1984; 20: 1089-1106 (in Russian).

[592]   Bolin B. The greenhouse effect, climatic changes and ecosystems.SCOPE 29. New York: John Wiley and Sons 1986.

[593]   Budyko MI. Man's influence on climate. Leningrad: Gidrometeoizdat 1972 (in Russian).

[594]   Kellogg W. Review of mankind's impact on global climate. In: Multidisciplinary research related to the atmospheric sciences: Boulder 1978, pp. 64-81.

[595]   Budyko MI. The Earth's climate: past and future. Academy Press: New York 1982.

[596]   The impact of atmospheric carbon dioxide increasing on climate. Proceedings of the Soviet-American Workshop on Atmospheric Carbon Dioxide Increasing Study, Leningrad, 15-20 June 1981. Leningrad: Gidrometeoizdat; 1982 (in Russian).

[597]   Hansen J, Rind D, Lacis A, *et al.* Global climate changes as forecast by Goddard Iinstitute for space studies three-dimensional model. J Geophys Res 1988; 93: 9341-9364.

[598]   Caillon N, Severinghaus JP, Jouzel J, *et al.* Timing of atmospheric CO_2 and Antarctic temperature changes across Termination III. Science 2003; 299: 1728-1731.

[599]   Monnin E, Indermühle A, Dällenbach A, *et al.* Atmospheric CO_2 concentrations over the last termination. *Science* 2001; *291*: 112-114.

[600]   Mudelsee M. The phase relations among atmospheric CO_2 content, temperature and global ice volume over the past 420 ka. Quat SciRev 2001; 20: 583-589.

Index

A

Accelerator mass-spectrometry (AMS) 63
ACRIM series 20, 21, 131
Activity, cyclonic 87, 145, 147
Aerosol concentration 140-142, 168, 193, 198
   sulphate 168
Aerosol data, used 141, 142
Aerosol layers 136, 140, 160, 161, 174, 178, 181
   new 140
   stratospheric 159, 160
Aerosol particles 86, 140, 142, 152, 157, 159, 175, 176, 178, 179
   concentration of 153
Aerosols 6, 60, 61, 139, 141, 143, 154, 157, 159-161, 167-169, 176, 180, 181, 192, 193, 207
Air, cleared humidified 175
Air bubbles 110
Air masses 83, 88
Albedo, global planetary 128, 129
Amplitude-period effect 23
Amplitudes of cycles correlate 23
Antarctica, internal areas of 87, 88
Antarctic atmosphere 141, 142
Antarctic Ocean 146, 147
Anthropogenic 8, 69, 89, 119, 127, 193, 206
Anticorrelation 23
Anticyclones 6, 87, 130, 147
ARS effect 119, 125, 189
Astronomers 14, 16, 92
Astronomic temperature cycles 116
Astronomic units 5
Atmosphere
   lower 30, 135, 136, 153, 176, 180, 205
   upper 6, 28
Atmosphere concentration 87
Atmosphere-ocean interaction 190
Atmospheric, global 56, 154
Atmospheric aerosols 139, 141
Atmospheric circulation 83, 134, 139, 151, 152, 154, 155, 182, 183
   large-scale 149, 183
   response of 145, 153
Atmospheric concentrations 68, 207
Atmospheric gases 28, 84, 110, 159
Atmospheric instabilities 6
Atmospheric ionization 85, 139, 153, 154, 166, 170
Atmospheric parameters 143, 171
Atmospheric pressure 147, 151, 153, 175, 182, 183
Atmospheric pressure variations 149
Atmospheric processes 6, 129, 130, 133, 145, 147

Atmospheric radiocarbon 59
Atmospheric response 142, 151
Atmospheric temperature 187
Atmospheric transparency 133, 136, 137, 139, 153, 156, 160, 167, 180, 181
Atmospheric turbidity 136, 137
Atmospheric weather 6
Atmospheric weather conditions 174
Aurora borealis 10, 28, 29
Auroral radiation 25, 26, 28, 30
Auroral zone 27, 134, 136, 152, 153
Axis, equatorial trough 150

B

Balance
   soil water 99, 100
   thermal 76
Beryllium concentration 62, 63, 78, 79, 81, 83
Beryllium records 63, 79
Black thick line 203
Blinova index 134-136
Borehole temperature 107, 112
Brightening, global 196, 197

C

Carbon, migration of 58
Carbon cycle, global 56, 58
Carbon dioxide 69, 201, 203, 207
Carbon dioxide concentration 193, 201, 203, 207
Century-long cyclicity 125, 167
Century-scale 42, 91, 95, 112
Century-scale climatic variability 155
Century-scale cycle 22, 95, 155, 166
Century-scale cycle band 90
Century-scale nitrate cycle 164
Century scale variations 74
Century sunspots 24
Chromosphere 3, 4, 11
Chronology 103, 116
Circuit, global 154
Circulation forms 147, 149
Climate, regional 64, 70
Climate changes 184, 185, 200
Climate evolution 125, 200, 202, 206
Climate history 113, 203
Climate modelling 190, 194
Climate models 190
Climate proxies 98, 107
Climate research 51, 98
Climate sensitivity 192, 198